T0234479

Dynamic Simulation and Virtual Reality in Hydrology and Water Resources Management

Dynamic Simulation and Virtual Reality in Hydrology and Water Resources Management

Ramesh S. V. Teegavarapu
Chandramouli V. Chandramouli

CRC Press
Taylor & Francis Group
Boca Raton London New York

CRC Press is an imprint of the
Taylor & Francis Group, an **informa** business

First edition published 2021 by
CRC Press
6000 Broken Sound Parkway NW, Suite 300, Boca Raton, FL 33487-2742

and by

CRC Press
2 Park Square, Milton Park, Abingdon, Oxon, OX14 4RN

Library of Congress Cataloging-in-Publication Data

Names: Teegavarapu, Ramesh S. V., 1970- editor. | Chandramouli, Chandramouli V., editor.
Title: Dynamic simulation and virtual reality in hydrology and water resources management / [edited by] Ramesh Teegavarapu, Chandramouli V. Chandramouli.
Description: First edition. | Boca Raton, FL : CRC Press, 2021. | Includes bibliographical references and index.
Identifiers: LCCN 2021004512 (print) | LCCN 2021004513 (ebook) | ISBN 9780367363789 (hardback) | ISBN 9780367363789 (paperback) | ISBN 9780429345555 (ebook)
Subjects: LCSH: Watershed management--Data processing. | Water resources development--Data processing. | Hydrologic models. | Virtual reality.
Classification: LCC TC413 .D96 2021 (print) | LCC TC413 (ebook) | DDC 627.01/13--dc23
LC record available at https://lccn.loc.gov/2021004512
LC ebook record available at https://lccn.loc.gov/2021004513

ISBN: 978-0-367-36378-9 (hbk)
ISBN: 978-1-032-04325-8 (pbk)
ISBN: 978-0-429-34555-5 (ebk)

Typeset in Sabon
by Deanta Global Publishing Services, Chennai, India

Contents

3 System Dynamics Models: Applications   99

Preface

Descriptive and prescriptive approaches are generally used for modeling and management of natural or engineered systems. The former is essential for understanding the system, while the latter will help in the improvement of any measure that quantifies the system performance. Simulation models as one of the best descriptive approaches are developed, calibrated, and validated for modeling and managing systems. Systems thinking helps to evaluate systems with different elements by evaluating interrelationships among these elements. Built on the foundations of systems thinking is the unique paradigm of system dynamics (SD), which helps in understanding the natural and engineering systems by considering the overall behavior of the system over time and interactions among different elements of the system. Material and information feedbacks (positive and negative) are used to explain, model, and predict the system behavior. Emerging approaches that use simulation with animation and virtual reality are beneficial to understand natural and engineering systems that are operated and managed by humans. This book focuses on the use of an SD-based simulation approach, simulation with animation, and virtual reality tools for modeling and management of hydro-environmental systems.

Stocks and flows that form the main elements of the SD approach are ideal for modeling water and environmental systems which deal with storage and movement of water in space and time considering quantity and quality issues. Specially designed object-oriented simulation environments that provide drag-and-drop objects with generic properties are used for the development of the SD models. The first three chapters of this book discuss the principles of the SD approach, building models for water and environmental systems, and case study applications using this approach. Modeling environments used for the development of SD models are also elaborated. Simulation along with animation in space and time allows visualizing the changes in the system in a dynamic interactive mode. The animation mode allows users to evaluate the systems under different scenarios that are otherwise not possible in static simulation. Chapter 4 describes the concepts related to simulation combined with the animation approach and documents an application of the same to visualize and assess the damages caused by a catastrophic flood. The approach uses geospatial and spatial flood inundation information obtained from the execution of a hydrologic simulation model. Virtual reality approaches and tools have received enormous attention in the past decade in engineering applications for a better understanding of the systems with the help of immersive environments. System simulation using these environments also add to improved pedagogical experiences and help policymakers to devise and analyze different scenarios aimed at decision-making. Chapters 5 and 6 provide a comprehensive introduction to virtual reality modeling approaches and tools and applications of these tools for modeling and management of water-related disasters (e.g., floods). This book is expected to serve as a valuable reference and resource for professionals, researchers, and students who are interested in developing simulation models with or without animation, exploring SD principles and applications, and creating virtual reality models for modeling hydro-environmental systems.

In closing, sincere thanks are extended to Tony Moore and Frazer Merritt at CRC Press and Routledge for all the help and support throughout the publication process.

Ramesh S. V. Teegavarapu
Chandramouli V. Chandramouli
November 2020

About the Authors

Ramesh S. V. Teegavarapu (Dr. T.) is a professor, Fulbright scholar, and graduate program director in the Department of Civil, Environmental, and Geomatics Engineering and founder of the Hydrosystems Research Laboratory (HRL) at Florida Atlantic University, Boca Raton, Florida. His main research interests include climate change and variability, hydro-climatic extremes, water and environmental systems modeling and management, spatial hydrology, and hydrometeorology. Dr. T is an author, co-editor, and sole editor of several books on these topics.

Chandramouli V. Chandramouli is a professor at Purdue University Northwest, Indiana, where he teaches hydrology and hydraulics courses. His expertise is in water resources and environmental modeling integrating artificial intelligence techniques, and his focus areas include reservoir operation and climate studies, flood modeling, and regional water quality analysis. He collaborated with the Center for Innovation and Visualization Center at Purdue University Northwest and developed virtual 3D lab modules for flooding in a real-world system, as well as groundwater contamination movement for classroom teaching.

Acronyms

3D	Three Dimensions (includes all the three major axes in a Cartesian coordinate system)
4D	Four Dimensions (includes 3D and also the time dimension)
ABM	Agent-Based Modeling
ACD	Activity Cycle Diagram
AHP	Analytic Hierarchy Process
AI	Artificial Intelligence
AMC	Antecedent Moisture Condition
AR	Augmented Reality
BASINS	Better Assessment Science Integrating point and Non-point Sources
BN	Bayesian Network
CADSWES	Center for Advanced Decision Support for Water and Environmental Systems
CAVE	Cave Automatic Virtual Environment
CFD	Computational Fluid Dynamics
CIVS	Center for Innovation through Visualization and Simulation
CLD	Causal Loop Diagram
CN	Curve Number
DEM	Digital Elevation Model
DES	Discrete-Event Simulation
DLL	Double-Loop Learning
DO	Dissolved Oxygen
DOD	Department of Defense
DOQQ	Digital Ortho Quarter Quadrangle
EMA	Exploratory Modeling and Analysis
EMB	Equation-based Modeling
FBX	Filmbox
FPS	Frames per Second
GFD	Graphical Functional Diagram
GIS	Geographical Information System
GT	Game Theory
GUI	Graphical User Interface

HEC-HMS	Hydrologic Engineering Center – Hydrologic Modeling System
HEC-RAS	Hydrologic Engineering Center – River Analysis System
HITL	Human in the Loop
HMD	Head-Mounted Display
HSPF	Hydrologic Simulation Program Fortran
IA	Impact Analysis
LH	Latin Hypercube
LHS	Latin Hypercube Sampling
LOD	Level of Detail
M & S	Modeling and Simulation
MAE	Mean Absolute Error
MC	Monte Carlo
ME	Mean Error
MRE	Mean Relative Error
MSE	Mean Squared Error
ODE	Ordinary Differential Equations
OOS	Object-Oriented Simulation
RAM	Random Access Memory
RBP	Reference Behavior Pattern
RK	Runge–Kutta
RK2	Runge–Kutta second order
RK4	Runge–Kutta fourth order
RMSE	Root Mean Squared Error
SD	System Dynamics
SDEP	System Dynamics in Education Project
SOP	Standard Operating Policies
TMDL	Total Maximum Daily Load
UI	User Interface
USDA	United States Department of Agriculture
USEPA	United States Environmental Protection Agency
USGS	United States Geological Survey
VPL	Virtual Programming Language
VR	Virtual Reality
WEAP	Water Evaluation And Planning
WGS	World Geodetic System

Chapter 1

Simulation and Systems Dynamics Approach

Ramesh S. V. Teegavarapu

1.1 INTRODUCTION

Understanding any physical, social, business, or economic system requires a model or observations related to one or more variables that define the status of that system at any point in time or space. Models as abstractions of reality are developed for understanding, analyzing, and evaluating physical systems. Mathematical models are developed to characterize the processes affecting the system behavior, and these models use analytical solutions wherever possible or simulation to analyze the system. Not all models can characterize the reality accurately, as the statistician George Box suggests, "All models are wrong, some are useful." However, models are useful if they are beneficial for simulation and optimization of the systems for better control and management. Models are used for prediction as well as forecasting under different situations. The focus of this book is mainly about models developed for understanding and modeling water resources and environmental systems, and therefore the rest of the book will focus on simulation models.

A model can be conceptual, physical, empirical, deterministic, or stochastic. Conceptual models are developed based on the knowledge of the physical systems and obey the laws of physics, which explain or describe the processes. An example of a simple conceptual model can be a mass balance model of a lake or a reservoir. The change in storage as defined by water level or volume in a lake or reservoir can be obtained mathematically using inflows and outflows in a time interval. For example, a well-known water balance model referred to as the ABCD model (Thomas, 1981) is a four-parameter (viz., A, B, C, and D) model, which is useful for water balance or water budget applications. The four parameters (A, B, C, and D) help quantify the runoff and rechange, saturation levels, groundwater rechange, and rate of groundwater discharge in a water balance model that contains surface and subsurface components. The parameters can be calibrated using a nonlinear optimization approach. Conceptual models have parameters that can be estimated through the calibration process and then validated for applications. A physical model is a representation of any natural or engineered system constructed or fabricated in a laboratory to a specific physical scale to understand or simulate a process. An engineered system in this chapter and book is referred to as any system that is human controlled or managed. Scaled models require dimensionality analysis to represent large-scale processes at a smaller level accurately. Empirical models are based on the concept of inductive modeling techniques and are based on observed datasets related to a process. These models are referred to as black-box models as they do not always provide functional or mathematical relationships between inputs and outputs of the model that can be easily understood. Even if appropriate functional forms are established between inputs and outputs, these relationships are sometimes too complex to understand the processes that are being modeled. Variants of empirical models may incorporate physically based parameters, and these models are referred to as semi-empirical models or approaches. Deterministic models assume that all the inputs and the parameters of the model are known with certainty and are not random. Stochastic models, on the other hand, assume that both inputs and parameters are uncertain.

The department of defense (DOD, 2011) of the U.S. in the Modeling and Simulation (M & S) glossary defines modeling and simulation as "The use of models, including emulators, prototypes, simulators, and stimulators, either statically or over time, to develop data as a basis for making managerial or technical decisions." In many studies modeling and simulation are used interchangeably to suggest the use of model or execution of the model for understanding, predicting, or forecasting the system behavior or variables influencing the system. Simulation can be used to address analysis, design, and control problems.

The use of simulation models and systems' thinking approaches is common in many disciplines. In the hydrologic modeling field, conceptual, physical, lumped (or spatially averaged), distributed models, time series, and data-driven models are used for different purposes, from hydrologic design to management. Lumped models use spatially averaged inputs and do not consider the spatial variability of inputs and are generally used for design purposes. The heterogeneity of the system in the spatial domain is not considered in these models. Distributed simulation models consider the variability in inputs by discretizing the spatial domain into smaller units referred to as tessellation and adopting inputs from each of these units. Time series models are generally used for forecasting or modeling observations at multiple temporal scales. The relationships between data at different time intervals are identified and are used to develop models for predictions. Data-driven models use data related to one or more variables to develop empirical or inductive models. These models are useful for a variety of applications, with a major limitation being their inability to explain the physical processes. Narrative models provide descriptions of the systems in words, and their use is limited to some fields such as computer engineering. Other types of models that may not involve computer simulation include normative and physical or prototype models. Normative models use the most commonly available materials and objects that are used by humans to explain the intricacies of complex systems and their workings. An analytical model represents a real-world system by a set of equations that are solved.

Simulation of any physical system requires the development of complete knowledge of the system with all the interrelated components of the system which interact with each in time. Modeling and assessment of system behavior can be carried out using discrete and continuous simulation models. The former evaluates the state of the system or state variables at specific points in time, whereas the latter assesses the system that changes continuously over time. It should be noted that the time interval at which discrete-event simulation (DES) is carried out need not be constant. Some simulation models combine both the aspects and characterization of discrete and continuous events. Discrete-event simulation considers events as well as variables which include time, counter, and system state (Ross, 2013). More discussion about modeling and simulation can be found in the works of Rossetti (2016), Birta and Arbez (2019), and Sokolowski and Banks (2009).

1.2 DESCRIPTIVE AND PRESCRIPTIVE MODELS

Descriptive models are those which help to describe the systems, and these models are appropriate for understanding and evaluating the system. These types of models are mainly simulation models, although prototype physical models can come under this category of modeling approaches. Prescriptive models belong to the category of optimization models that maximize or minimize an objective that quantifies the system performance. Optimization models attempt to improve system performance. The optimal magnitudes of decision variables are primarily obtained by the application of mathematical programming or optimization techniques and with appropriate solution techniques or approaches (i.e., solvers). In many instances, simulation and optimization models are used in tandem while the former provides a quantification of system behavior or performance and the latter optimizes the system performance using single or multiple objectives.

1.3 SYSTEMS AND SYSTEMS THINKING

A system is defined as a set of interconnected parts or components that form a whole. A well-defined structure or organization of the parts can be identified in a system with variables influencing the state of the system. The relationships between the variables can be established using functional relationships. A system may consist of one or more sub-systems, and the processes at each sub-system collectively influence the overall behavior of the system. The systems thinking paradigm involves identifying the elements of the system and interactions between these elements to effectively understand the interactions to evaluate the performance of the system over time and in space. The reader is referred to excellent introductory books on systems thinking by Maani and Cavana (2000), Forrester (1968), Kauffman (1980), Meadows (2008), Kim (2000), Sweeney and Meadows (2010), and Rutherford (2020). Excellent discussions on the need for systems thinking skills have been provided by Sterman (2002), Richmond (1993), and Sherwood (2002). Forrester (1994) provided a comparative analysis of systems thinking, system dynamics (SD), and soft operations research (OR) paradigms. On those similar lines, Kunic (2018) compiled an extensive set of works discussing the links between SD and soft and hard OR. In the case of general, soft OR deals with qualitative analysis and hard OR are based more on a quantitative analysis involving mathematical models. Eisner (2019) and Mella (2007) provide a comprehensive review of approaches along with systems thinking which include design, lateral, disruptive, critical, inductive, deductive, out-of-the-box, fast and slow, and reductionist thinking. Applications of systems thinking and soft operations research concepts to manage complexity in different types of systems are discussed by Masys (2016).

1.4 DYNAMIC SIMULATION

Dynamic simulation refers to an explicit process of evaluation of system changes in space and time. Sometimes models developed for evaluating changes in a system over time are referred to as dynamic models. Once the spatial extent of the process to be modeled is identified and fixed, changes over time are critical for predicting the system behavior in the future. The stationary or non-stationary nature of the system depends on the inputs to the system and variables that define the behavior of the system. The dynamic simulation should not be confused with the modeling paradigm or field referred to as dynamic systems. The reader is directed to excellent reviews of different systems simulation and modeling paradigms by Borshchev and Filippov (2004) and books by Osais (2018), Gordon and Guilfoos (2017), and Birta and Arbez (2019).

1.4.1 Dynamic Simulation Environments

Dynamic simulation environments designed and built to understand any system are better than the spreadsheet-based modeling/simulation environments. Simulation using a spreadsheet, or any computational code, is restrictive in several ways to understand the behavior of any system. Any dynamic simulation environment is expected to have the following features:

- A visual depiction of the complete system with different components of the system and their interactions shown by connections or links.

- Scenario generation using simulation runs with varying inputs and almost instantaneous results.
- Capability to visually show the feedbacks or interactions between different components within the system.
- Clarity of the quantitative expressions or equations used to evaluate the system behavior in time and space.
- Separation of data and results and focus on overall behavior than numerical values.
- Provides a user interface (UI) with the capability of multi-user model development.
- Availability of checks on model structure and validity.
- In-built feature to assess parameter and output uncertainty through sensitivity analysis and advanced features such as Monte Carlo (MC) simulation or MC with Latin Hypercube (LH) sampling.
- Ability to calibrate the model through manual and automatic procedures with the latter relying on optimization methods.
- Ability to conduct checks of dimensions, gain insights into interactions between variables and receive information about variables.
- Visual depictions of changes in variable values and different execution speeds and graphical user interface (GUI) for presentation of results.
- Ability to build reliable and robust simulation models at a rapid pace.
- Simulation at different speeds to monitor the behavior of the system.
- User interface (UI) to modify the parameter values of the systems for analysis of scenarios.
- Ability to see the model components at different levels of detail.
- Separate qualitative and quantitative information for decision-makers.
- Ability to share the models to other uses without having to have the proprietary software.
- Ability to import and export data from the simulation environment.

Users of the simulation environments expect most of these features to be included in their modeling environments.

1.5 SIMULATION ENVIRONMENTS SPECIFIC TO HYDROLOGIC MODELING AND WATER RESOURCES MANAGEMENT

Several simulation environments are developed in the past four decades for the evaluation of the water and environmental systems. There are two categories of simulation environments that are used for modeling water or environmental systems. The first category of simulation environment is generic that provides the model developers generic objects or computational codes available under a specific platform to model the system under consideration. The generic objects can be used with or without modification and attach any additional specific properties to model different elements of the system. The second category of simulation environment will provide modeling objects that already have specific properties assigned and can be used in restrictive environments for specific problems. Examples of such environments include HEC-HMS (Hydrologic Engineering Center – Hydrologic Modeling System), HEC-RAS (Hydrologic Engineering Center – River Analysis System), WEAP (Water Evaluation and Planning), and Riverware developed by the Center for Advanced Decision Support for Water and Environmental Systems (CADSWES) of the University of Colorado, Boulder, USA, and several others developed by both public and private entities. More general and easy performance evaluation of water resources systems can be carried out by developing models using object-oriented simulation (OOS) environments. The *Simulink* graphical modeling environment supported by MATLAB that uses *block* and *arrow* diagrams is also suitable for simulating water and environmental systems. A suite of mathematical and logical functions and visualization capabilities with specialized toolboxes provide users with the ability to model any complex system and simulate the model behavior over time. *Simscape*, an extension of *Simulink* can be used for modeling different physical systems. For example, *SimHydraulics* can be used for modeling and simulating hydraulic systems (Esfandiari and Lu, 2018). The *GoldSim* is another software that provides a visual spreadsheet with tools for simulating and visualizing complex science, business, and engineering systems.

1.6 SYSTEM DYNAMICS APPROACH AND SIMULATION

System Dynamics (SD) (Forrester, 1958, 1961; Wolstenholme, 1990; Coyle, 1996; Richardson, 2011; Dangerfield, 2013; Keating, 2020) is a concept based on systems thinking where dynamic interaction between the elements of the system is considered to study the behavior of the system as one entity. As the name suggests, the behavior of the system is monitored over time and therefore dynamic as opposed to static. SD approach should not be confused with the concept of dynamic simulation. A snapshot of the system can be obtained in any dynamic system by identifying the status of the system at a specific time interval. The concepts of SD were initially introduced by Forrester (1958, 1961). The main idea of SD modeling is to understand the behavior of the system using simple mathematical structures. SD concepts are aimed at understanding the time-dependent behavior of complex systems. The models built using SD concepts are descriptive that explain the interaction between different elements of any system. SD concepts can help (i) describe the system; (ii) understand the system; (iii) develop quantitative and qualitative models; (iv) understand how information feedback (Åström and Murray, 2008; Martin, 1997) and delays govern the behavior

of the system and finally; and (v) develop control policies for better operation or management of the systems. Initial system description and a basic
understanding of the structure of the same is achieved by the development
of mental models. These models are based on one's thinking about how a
real-world system works and sometimes they can be vague, biased, ambigu-
ous, and non-testable as pointed out by Barlas (2002). One of the main
advantages of system dynamics is that the models can be developed for
understanding the behavior of the system using simple building blocks and
they are causal-descriptive (Barlas, 1996) and so are called white-box models (Duggan, 2016). The SD approach deals with system behavior at the
macro-level or referred to as an aggregate-level modeling as opposed to an
individual-level (or element level) evaluation. SD is a concept whereas the
latter refers to a simulation that focuses on the evaluation of any system
with time or space. The SD approach is suitable when:

- Aggregate system behavior with time needs to be understood and not
  necessarily the changes in the individual components of the system.
- Different feedbacks influence the main variables in the system through
  information and material flows.
- Minor or major changes to variables at different points of time might
  influence the status of the system over time.

SD approach is beneficial for systems with elements that show interdepen-
dence, have mutual interaction and information feedback, and are affected
by circular causality (Richardson, 1996; SD, 2020). Circular causality, a
well-known concept in the field of cybernetics, refers to a two-way interaction between two elements of the system, with both the elements initiating
the cause and bear the effect of a specific interaction. Circular causality is
more complex than linear causality, where there is only one-way interaction
between any two elements of the system. Hester and Adams (2014) indicate that SD operationalizes the concepts of cybernetics with communication and control to model the behaviors of complex systems. SD approach
can also help in addressing the dynamic complexity of the system, which
stems from changing cause and effect relationships over time. This complexity in systems, according to Sterman (2001), is mainly attributed to
characteristics such as: (1) constantly varying nature; (2) tightly coupled or
connected; (3) governed by feedback; (4) nonlinear; (5) history dependent;
(6) self-organizing; (7) adaptive; (8) counter-intuitive; and (9) policy resis-
tant. Systems that change over time are dynamic, and those that are tightly
coupled have elements within them that interact and influence the behavior
of the systems. Adaptive systems evolve and re-adjust according to new
or gained knowledge or interventions brought by the elements of the system of humans. Some systems resist and do not align or converge to target
behavior or state and become what is known as policy-resistant systems.
Persistence in some systems plays a role in systems becoming dependent on

past behaviors. Counteractive behavior is common in many social systems where unusual states of the system are noted that are not expected based on specific policy implementation. Dynamic complexity is difficult to assess when compared to detail complexity that relates to the existence of many variables and the interrelationships between them in a static model. SD also promotes closed-loop thinking as opposed to the open-loop thinking process. The latter tries to address any problem without utilizing the concepts of feedbacks. SD modeling is also referred to as equation-based modeling (EBM) as it involves the development of a set of equations that characterize the system over time using ordinary differential equations (ODEs).

According to Sterman (2000):

> System dynamics is a perspective and set of conceptual tools that enable us to understand the structure and dynamics of complex systems. System dynamics is also a rigorous modeling method that enables us to build formal computer simulations of complex systems and use them to design more effective policies and organizations. Together, these tools allow us to create management flight simulators-microworlds where space and time can be compressed and slowed so we can experience the long-term side effects of decisions, speed learning, develop our understanding of complex systems, and design structures and strategies for greater success.

The readers are referred to several books related to general SD concepts and applications in different fields in the last three decades. General discussion about the SD approach along with general applications can be found in the books by Coyle (1996), Moffat (1991), Roberts et al. (1994), Ghosh (2017), Forrester (2009), Duggan (2016), Bala (1999), Bala et al. (2017), and Goodman (1989). Applications of SD for environmental modeling are discussed by Belt and Dietz (2004), Ford (1999), Nirmalakhandan (2002), and Deaton and Winebrake (2000). SD model development and applications can be found in the works of Teegavarapu and Simonovic (2014), Simonovic (2010), and Simonovic (2008). Applications of SD principles and approaches for economic and business operations are elaborated in books by Sterman (2000), Ruth and Hannon (2012), Campuzano and Mula (2011), Morecroft (2015), and Garcia (2019). Modeling of diseases and pests using SD modeling approaches is discussed by Ruth and Hannon (2009) and community and group-based model building methodologies are elaborated by Hovmand (2014), Vennix (1999), and Richardson and Andersen (2010). Applications of SD in health care are discussed by Wolstenholme and McKelvie (2019). Comparative evaluation of discrete-event simulation and SD approaches is provided by Brailsford et al. (2014). Applications of SD models for improved learner-centered education for teaching water resources systems in a classroom setting are briefly discussed by Teegavarapu (2018a). The system dynamics society website (www.systemdynamics.org) and its journal, *System Dynamics Review*, are two excellent sources of information about the current state-of-the-art development and applications of SD models. The website https://thesystemsthinker.com/ provides information about SD principles and systems thinking. A concise SD glossary developed by Ford (2018) is an excellent reference for readers. The concepts of SD are briefly described in this chapter. A brief discussion about the development of SD models is provided in the next few sections. Building and applications of the SD models for water and environmental systems will be discussed in Chapters 2 and 3, respectively.

The SD approach of building models of systems is summarized by the following steps:

- Identify and define the problem using systems thinking principles with a specific spatial and temporal scale.
- Collect all the data necessary to characterize the problem as a dynamic one that requires considerations of changing variable values with time.
- Identify all the variables and specify the structure of the system looking inward first (endogenous point of view or looking within the boundary of the system under consideration).
- Develop an initial model (i.e., mental model) to identify relationships between different variables and causal connections (feedback loops).
- Develop a computer simulation model and simulate the behavior of the system with an initial evaluation of changes in different variables over time.
- Calibrate and validate the model with available data to assess its utility in characterizing the realistic behavior of the system.
- Test the structure of the model to assess its capability in replicating the real system behavior in all possible conditions with data that is not used for calibration.
- Use the validated model to assess system behavior based on planned changes (or policies) in different variables and conditions of the system over time. Forecasting and backcasting of the system can be carried out.
- Work with modelers, stakeholders, and decision-makers to revise, improve, and re-calibrate and validate the model if necessary and to develop implementable policies to achieve the required results.

Morecroft (2015) summarizes all these steps in a more concise form utilizing iterative five stages, and they include: (1) articulate problem; (2) propose dynamic hypothesis; (3) build simulation model; (4) test simulation model; and (5) design and evaluate policy. The initial exercise of qualitative analysis of the system using causal loop diagrams and word-and-arrow diagrams to provide descriptive versions of the processes is reflected in the soft perspective in SD (Kunc, 2018). The hard perspective of the SD relates to quantitative analysis or simulation by modeling the system using stocks and flows for use in a prescriptive/predictive process (Kunc, 2018).

1.7 DEVELOPING SD MODELS

The development of SD models will involve two phases, qualitative and quantitative. The first phase involves the development of an understanding of the system, gaining knowledge about the system, describing the system, identifying variables, and defining relationships by cause and effect relationships or diagrams. This phase also involves the development of mental models (Doyle and Ford, 1998) which are preliminary ideas about the different elements of the system, interconnections among the elements, and a structure of the system with or without any understanding of the system behavior. The second phase involves the development of mathematical relationships, linking different elements of the system and evaluating the system states and behavior to make logical conclusions or plan for operational changes. The following sections describe the SD model development in the form of a series of steps.

1.7.1 System Identification

An understanding of the system to an extent to define the boundary of the system, variables that are critical in developing mathematical functional relationships, is also required for the development of simulation models based on the SD approach.

1.7.2 Causal Loop (Influence) Diagrams and Feedback

As the first step in the development of SD models, causal loop diagrams (CLDs) or influence diagrams (IDs) (Kim, 1992; Moffat, 1991; Garcia, 2018a, 2018b; Cavana and Mares, 2004; Roberts et al., 1994) are developed for representing the interrelationships between various elements of the system. They also help to define the cause–effect links (or causal links) and form closed loops. All CLDs provide information about feedbacks in the system. Feedback in simple terms is the transfer of information from one variable to another variable to characterize the influence of the former on the other. Richardson (1999), Maani and Cavana (2000), and Cavana and Mares (2004) provide a good discussion about feedbacks. Feedbacks can be positive and negative with the former having a reinforcing effect and the latter having a balancing effect.

The following steps are recommended for the development of CLDs:

- Identify the variables that are known to influence system behavior.
- Use arrows to link one variable to another to define cause and effect relationships. These arrows (or directional links), when associated or connected with variables, form one or more loops.
- Indicate the polarity by using signs ("+" or "–")at the end of each arrow to show how one variable influences the other.

- Identify any delays in the system and mark them in the diagram.
- Assign the polarity for each loop if multiple loops exist. A reinforcing loop is identified by a "+" sign and a balancing loop with a "−" sign. These polarity signs are generally placed in a parenthesis.

All CLDs will require four components, and they are variables, links, polarity sign for each link, and a sign for loop indicating negative or positive feedback. A link is used to imply the direction of causation and not a time sequence of occurrences of events associated with the variables. The positive and negative polarities indicate changes in variables either in the same or opposite directions. It is not uncommon to find CLDs with polarity signs "+" and "−" for the links replaced by "S" for the same and "O" for opposite, respectively. Also, the signs for reinforcing and balancing loops (or negative feedback loops) are indicated by letters "R" and "B," respectively, in CLDs. Richardson (1986) provides an excellent discussion of the advantages and disadvantages of using polarity signs and "S" and "O" letters. Examples of causal loop diagrams for understanding the human population, carbon metabolism in plants, food security, fisheries, electricity supply, and global warming problems are provided by Bala et al. (2017). An exhaustive discussion about reinforcing and balancing feedback loops and CLDs is provided by Kirkwood (1998). The main limitation of CLDs is that they depict qualitative and not quantitative relationships. Therefore, CLDs are noted as diagramming or mapping tools that form the initial qualitative models of the SD approach. Guidance on the development of error-free and non-ambiguous CLDs is provided by Garcia (2018b) and Bureš et al. (2019). As a precursor to SD model development, CLD creation needs to be carried out with caution and the process should conform to the established notation for naming the variables as well as the fixed set of geometric shapes that can be used.

1.7.3 Building Blocks of SD

Central to the theme of systems dynamics (Forrester, 1961) are two important building blocks, stocks and flows, that can be used to model the elements of any system. Equally important are the connectors, converters, sinks, and sources. Stocks (or reservoirs) are accumulators; they collect material over time and their behavior is controlled by flows. Flows are the inputs to the stocks, and they can vary with time or can be constant. Stocks can be used to model both tangible and intangible quantities such as water in a reservoir and buildup of knowledge, respectively. Connectors act as live links that transfer quantitative information from one block to another. Convertors perform the essential function of holding values of variables, graphical functional diagrams (GFDs), conditional rules, and conversion factors for dimensional consistency. They are also referred to as auxiliary variables. GFDs help to characterize the relationship between any two variables. The sources and sinks in one way define the boundaries of the system with flows appearing from sources and disappearing into sinks. Fieguth (2017) describes three systems based on the boundaries that allow matter and energy transfer, and they are: (1) open system (with energy and matter transfer, combined with sources and sinks); (2) closed system (with energy transfer only); and (3) isolated system (with no energy or matter transfer). A stock can also store material for a specific amount of time (i.e., transit time) before allowing the material to flow out. In this situation, the stock is referred to as a conveyor. Conserved and non-conserved flows are possible when the amount of material transferred from one stock to another is conserved and unlimited material is possible from a source.

An example of a causal loop diagram for the human population without any resource (or capacity) constraints depicting positive feedback is shown in Figure 1.1a. The links in the causal loop diagrams should represent cause and effect relationships and not correlations. Flow and stock–flow diagrams are then extracted to develop simulation models. These diagrams are also referred to as structural diagrams, and they are similar to activity cycle diagrams (ACDs) used for model design specification in discrete-event simulation (DES). A flow diagram for the population system is shown in Figure 1.1b. Four elements (viz., stock, flow, connector, and

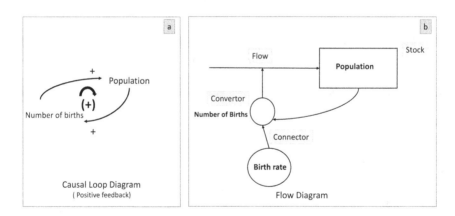

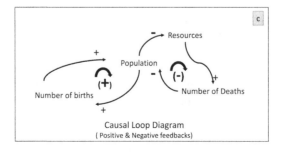

Figure 1.1 Causal loop and flow diagrams showing feedbacks.

converter) are used to create the flow diagram, and these are considered the building blocks of the SD model. A stock is an element that accumulates or drains over time. Flows help accumulate or drain the stock and connectors carry information from one element to another. The status (or magnitude) of the stock depends on the magnitudes of inflow and outflows. Stocks increase, decrease, or do not change when inflows are greater than outflows, outflows are greater than inflows, and outflows are equal to inflows, respectively. Convertors carry relationships or variable values that help link different elements of the system via connectors and do not accumulate or drain. The population is considered a stock that accumulates over time due to the flow of births. The births are estimated as a fraction of the population and are depicted using connector and convertor. Figure 1.1c shows the causal loop diagram that combines positive and negative feedback. Negative feedback (or compensating feedback) in this situation works as a balancing loop that constrains the population level by modeling deaths as a function of population levels. As the population increases, the number of deaths increases due to a lack of resources or any other constraints. A graphical function diagram (GFD) that relates two variables (viz., population level and death rate) can be created to characterize the limits to population growth. These limits are due to constraints on the resources (e.g., land and food), capacity (e.g., area), and others related to population dynamics as explained by Bacaër (2011).

1.7.4 Soft Variables

In some systems, the elements or the interrelationships between two or more elements are difficult to model as these elements or interrelationships are difficult to quantify numerically. Often the relationships are understood only in a qualitative way. For example, the willingness of a user and preferences of a modeler or a decision-maker toward an option specified verbally or using a qualitative scale can be modeled using soft variables. These variables are common in many socio-economic systems in which uncertain, imprecise, and vague information influence the decision-making and policy generation process. Operation and management of water and environmental systems are also dominated by situations that require engineering judgment, heuristics, historical knowledge, and preferences calibrated and developed based on past experiences. The SD approach allows the inclusion of soft variables by allowing the transformation of qualitative information to continuous numerical values to influence the system behavior over time. Flexible or user-modifiable graphical relationships between different variables allow the incorporation of soft variables in SD models. These graphical relationships are similar to fuzzy membership functions (Teegavarapu and Simonovic, 1999) that are mainly used for quantifying imprecise information. Soft variables provide mechanisms to model causative influences in systems.

1.8 UNDERSTANDING SYSTEM BEHAVIORS

The behavior of any system can be evaluated or monitored using one or more variables that influence the system. Modeling system behavior requires an understanding of how the system will change to minute modifications in one component of the system. Sometimes these changes can be counter-intuitive, and in data-rich environments, the behavior is difficult to model due to the increase in the number of variables that define the state of the system at any given time. Also, in some cases, the overall behavior of the system does not change with major modifications to one or more components of the system. Behaviors and archetypes (patterns of structure) of system variables are explained by Senge (1990) and Senge et al. (1994). Some of the behaviors discussed by Senge are (i) exponential growth or collapse/decay; (ii) movement toward a target value with or without delay; (iii) variation that alternates between two states (i.e., oscillatory or cyclical); and (iv) growth that levels off. These behaviors are characterized and depicted visually as exponential growth, J-shaped, and sigmoidal ("S" shaped) curves. Often these behaviors are influenced by environmental resistance or density-dependent carrying capacity in both natural and social-economic systems. Environmental resistance in ecologic studies refers to an aggregation of limiting factors (e.g., availability of food and the prevalence of predators) that constrain the exponential growth of individuals (e.g., organisms or plants). A J-shaped curve refers to sudden or abrupt cessation of exponential growth, mainly due to environmental resistance. A specific length of time associated with exponential growth is "doubling time" which denotes the time taken by the variable to double itself in amount. Oscillatory behaviors are often noticed in systems with predator–prey interactions and population dynamics of predator and prey animals in nature. The dynamics of these behaviors are explained by the well-known Lotka–Volterra model. Oscillatory behaviors are also found in production/distribution systems. Visualizing feedbacks sometimes is difficult without the mental models, causal loop diagrams, and final simulations based on an SD environment to assess the changes in the system over time. Descriptions of the system behaviors are provided in this section and are illustrated in Figure 1.2. In general, the order of the system is determined based on the number of stocks adopted to model the behavior of elements of the system.

Linear growth/decay: a system behavior (or a response of the system with time) can be due to constant flows. A linear growth or decay model can be developed using a single stock and a flow. A constant inflow (C) feeding into a stock (S_t) will result in linear growth of stock, while a constant outflow (D) from the stock will result in linear decay. In the latter case, a non-zero initial stock value is required. The linear growth and decay with constant inflow and outflow over time interval Δt are represented by Equations 1.1 and 1.2. The rate of change of stock with respect

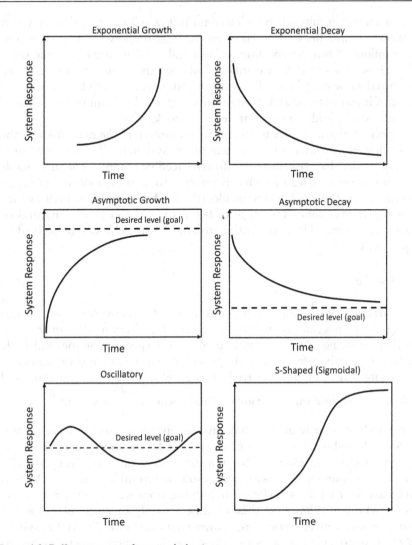

Figure 1.2 Different types of system behaviors.

to time is equal to a constant value (C for linear growth and –D for decay). Equation 1.3 represents the stock influenced by constant inflow and outflow simultaneously.

$$S_{t+1} = S_t + C\Delta t \quad \forall t \tag{1.1}$$

$$S_{t+1} = S_t - D\Delta t \quad \forall t \tag{1.2}$$

$$S_{t+1} = S_t + (C - D)\Delta t \quad \forall t \tag{1.3}$$

Linear growth results when the constant inflow (C) is larger than constant outflow (D), and decay is noted when constant outflow is larger than constant inflow. When the constant inflows and outflows are equal, the value of stock becomes zero. Linear inflow and constant outflow that is less than the initial value of inflow will lead to the nonlinear growth behavior of the stock. A linear outflow and constant inflow that is less than the initial value of outflow will lead to nonlinear decay of stock.

Exponential growth/decay: uncontrolled growth in the magnitude of the main element (variable) of the system or the system itself. This behavior is mainly guided by positive or reinforcing feedback and is not sustainable and can sometimes lead to what is referred to as overshoot and collapse. In economics, these behaviors are identified as boom and bust cycles which may repeat over time. This type of behavior is not evident in natural or managed systems. The exponential growth of a stock, S, can be defined by Equation 1.4.

$$S_t = S_o e^{kt} \tag{1.4}$$

The variable S_t is the value of the stock at time t, S_o is the initial stock value at $t = 0$, and k is a rate constant in Equation 1.1. A decay model can be developed by replacing e^{kt} with e^{-kt} in Equation 1.1. Depending on the model, the variable (k) can be referred to as the growth rate (fraction) or decay constant. The reciprocal of $k(-k)$ is referred to as growth (decay) time constant and inflow and outflow values for growth and decay are $\dfrac{dS}{dt} = kS$ and $\dfrac{dS}{dt} = -kS$, respectively. Exponential growth/decay contributes to system behavior that is undesirable and contributes to vicious cycles linked to reinforcing or amplifying processes (Road Maps, 2020). On the other hand, virtuous cycles with the same reinforcing processes may result in favorable outcomes. Vicious cycles are referred to as deviation amplifying loops with counter-productive results (Masuch, 1985). Both these cycles are mainly due to circular causality that suggests an effect becomes a causative factor for future effects (Korzybski, 1994). It is possible that a vicious cycle may be virtuous depending on the perspective with which the system is evaluated by a specific stakeholder involved in operational aspects of the system.

Asymptotic growth/decay: this type of behavior is mainly referred to as goal-seeking behavior, introduced by the negative feedback influencing the system. The response of the system approaches slowly toward a goal governed by balancing loops. In this context of goal-seeking behavior, a half-life is defined to indicate the time taken for stock to reach midway toward its intended final goal or state.

Oscillatory: a behavior associated with system response following above and below a specific average response. Generally, this behavior is due to the existence of higher order levels, and oscillations over time can be sustained (i.e., constant), expanding, and dampening, and chaotic. Oscillatory behavior is generally associated with delays in the system. Typically sustained, dampening, and expanding oscillations will have a constant, decreasing, and increasing amplitude over time, respectively. The chaotic oscillation has irregular amplitude over time. Delays in the systems may be caused by dead time (i.e., a finite amount of time before the system responds to change in the input) and lag (i.e., time is taken by the system completely to respond to the change in the input). A simple oscillatory behavior of a stock is noted when an oscillating inflow and constant outflow exists. Oscillatory behavior is generally noted in systems with two or more stocks (second and higher order) and negative feedback. Also, system behavior is influenced by two types of delays: (1) material and (2) information. In case of material delay, inflows or outflows can be delayed to and from stocks, thus causing a time lag in the growth of stocks and flows over time. Information delays also referred to as detection delays exist when there is a time lag due to changes in the way a specific variable in the system triggers changes in other variables. The term "relaxation time" from the field of cybernetics is also applicable in the case of a system that is disturbed. The relaxation time is the time a system takes to return to an equilibrium state or steady state after it experiences a disturbance. This time can also be linked to the resiliency of the system, although it relates to recovery after a failure (i.e., an undesired state of the system).

Goal-seeking: a behavior in which the system seeks the goal or approaches the pre-specified target with help of balancing loops governed by negative feedback loops in the system. The convergence to the target (or pre-defined goal) can be slow and an equilibrium point will be reached by the system. In some systems, this behavior can lead to self-regulation. In biological and ecological systems, this process is referred to as homeostasis. Adams and Mun (2005) refer to homeostasis as the property of an open system to regulate its internal environment to maintain a stable condition, utilizing multiple dynamic equilibrium adjustments controlled by interrelated regulation. Maintaining specific body temperature or blood sugar levels in humans is also attributed to homeostasis. The stability of the system attained during this time is identified as a temporary static equilibrium, a state in which minute changes will occur but the steady conditions prevail for the system. A variant of homeostasis state, in which the equilibrium achieved is dynamic, is referred to as homeorhesis.

An example of goal-seeking behavior in engineered systems will be a thermostat or regulator that controls temperature levels in a closed environment or setting. Another example of negative feedback in action is the working of the Watt governor, mainly developed for controlling engines run by steam. The goal-seeking behavior of a stock, S, can be modeled by Equation 1.5.

$$S_t = G_o \left(1 + \frac{S_o - G_o}{G_o} e^{-kt} \right) \tag{1.5}$$

The variable S_t is the value of the stock at time t, S_o is the initial stock value at $t = 0$, k is the decreasing rate constant, and G_o is the goal value that is being sought. The function value of S_t reaches G_o when the time approaches infinity. The function is referred to as bounded exponential, and Equation 1.5 can also be used to characterize asymptotic growth or decay behavior by defining appropriate goal value.

S-Shaped growth: a behavior that has an initial exponential growth and gradually approaches a maximum value governed by positive and negative feedback. The former feedback changes to the latter over time and this shift is referred to as loop dominance. The "S" shaped growth behavior (or logistic curve) is often noticed in systems that have a resource constraint that gets activated due to internal or external response to growth. Examples of such system behaviors are seen in animal and human populations constrained by space and the availability of resources for survival. This type of behavior is also referred to as density-dependent or constrained growth.

Steady state: the behavior of the system is constant over time. The system is considered to be in a stationary state.

The overall behavior of any system is mainly governed by dominant feedback in the system (negative or positive). An example of positive feedback is the uncontrolled growth of a species or population with no restrictions or constraints (e.g., space, food, water, and other essential resources). Positive feedback, also referred to as exacerbating feedback, occurs in many instances from simple audio feedback in sound systems with microphone and speaker or uncontrollable chain reactions. This feedback often leads to instability and needs counteracting negative feedbacks to stabilize the system. In many engineered systems, negative feedback or controlling or balancing feedback is commonly encountered and is an enforced mechanism to achieve stability of the system. Negative feedback helps the systems to be in a balanced mode. These systems are referred to as stable systems, and they exhibit goal-seeking behavior. Some of the feedback mechanisms are natural or self-induced (for example, in the case of sustainable ecosystems), and some others are human-controlled or engineered systems. In most of the managed short- or long-term operation of water resources systems, managers/operators enforce the feedback control with operation rules derived in consultation with stakeholders and decision-makers. These operation rules that are referred to as standard operating policies (SOP) are subject to revision based on changes in the system or due to the manifestation of extreme events.

External shocks are exogenous system variables that may have a sudden or unanticipated impact on the overall system behavior. These variables are realized outside of the system boundaries as opposed to endogenous variables that are mainly internal to the system. Ghosh (2017) lists several common generic system behaviors, and they are: (1) limits to growth; (2) shifting the burden; (3) drifting goals; (4) escalation; (5) success to successful; (6) tragedy of the commons; (7) fixes that fail; and (8) growth and underinvestment. These generic behaviors are elaborated by Bellinger (2020). Limits to growth is a classic capacity-constrained growth of natural systems. Shifting the burden refers to an attempt to correct a process that led to specific behavior of the system without considering the main factors (or variables) that are responsible for such behavior. System behavior is adjusted over time, leading to drifting goal state as information that becomes available based on simulation suggests that initially defined target is not achievable. Escalation is mainly a positive feedback behavior that leads to a vicious cycle. In the case of success to successful, one system behavior is more favored than the other depending on the system performance. The tragedy of the commons refers to the overexploitation of a common resource that results in depletion of the resource over time. Sometimes counter-intuitive behavior (Forrester, 1971a, 1995) is noted when expected results are not realized when a system evolves with time. This behavior can be associated with the failure of interventions to system behavior to obtain a specific target. Counter-intuitive behavior is possible due to the presence of aleatory and epistemic uncertainty associated with one or more elements of the system. The former type of uncertainty is mainly due to randomness within the system, and the latter is associated with the lack of knowledge about the processes influencing the system. Growth and underinvestment refer to business-specific behavior where resources are spent more in one sector (unit) compared to others that are underperforming.

Meadows et al. (2006) provide an excellent treatise on four different behaviors noticed in systems: (1) continuous growth (or exponential growth); (2) sigmoidal (or sustainable) growth; (3) overshoot and oscillation; and (4) overshoot and collapse. Reinforcing feedback and a combination of reinforcing and balancing feedbacks are noted in exponential and sigmoidal growth behavior, respectively. A growth with overshoot behavior is mainly due to reinforcing and balancing feedbacks with delay. A sustained oscillation is due to balancing feedback with delay. Overshoot and collapse behaviors are noted with reinforcing, balancing, and then a reinforcing behavior in that order. The directions of the first and the last reinforcing behaviors may be different. SD model simulations may also result in two different ending states based on the initial states of the system referred to as equifinality and multifinality. In the former case, the final state reached is the same irrespective of initial states and, in the latter, the ending states are different even though the initial state is the same. It is always important to assess the reference behavior pattern (RBP) or basic changes in the system over time. Instead of just looking at episodic events, a decision can be made regarding the time horizon of simulation, generic and main patterns in the system behavior using RBP.

Amadei (2015) explains generic structures possible using single and multiple stocks and flows and they include: (1) constant growth rate (stock and inflow and no outflow); (2) constant decay rate (stock and outflow and no outflow); (3) growth and decay (stock, inflow, and outflow); (4) exponential growth compounding process (stock, inflow, converter (growth rate constant), and no outflow); (5) compounding process and constant decay rate (stock, inflow, converter (growth rate constant), and constant outflow); (6) exponential decay draining process (stock, outflow, converter (decay rate constant), and no inflow); (7) draining process with constant growth rate (stock, constant inflow, outflow, and converter (decay rate constant)); (8) compounding process and draining process (inflow, outflow, stock, and converters (growth and decay rate constants)); (9) stock adjustment process and goal seeking (two stocks (one of them defining the stock), two converters, and inflow as bi-flow); (10) logistical S-shape model carrying capacity (two stocks (one of defining the capacity), two converters, and inflow); (11) compounding and draining process with variable rates (stock, inflow, outflow, constant growth, and varying decay rate); (12) draining and compounding processes with variable rate (stock, inflow, outflow, varying growth rate, and constant decay rate); (13) two interacting populations with possible oscillatory behavior (stock and inflow and outflow with growth and decay rates for each population); (14) transmission model (two stocks and converter (suggesting transfer rate from one stock to another); (15) overshoot and collapse (two stocks and inflows and outflows characterizing growth and resource availability constraints); (16) conservative chains (two stocks and inflows and outflows with material moving from one stock to another is conserved). A few generic behaviors listed here are discussed with simulations using an SD modeling environment. The readers are recommended to evaluate each generic behavior listed using a simulation environment to gain a comprehensive understanding of system behavior due to different inflow, outflow, and stock combinations.

1.9 TESTING AND VALIDATION OF SD MODELS

According to DOD (2011), verification refers to the process of determining that a model or simulation implementation accurately represents the developer's conceptual description and specification and validation refers to the process of determining the degree to which a model or simulation and its associated data are an accurate representation of the real world from the perspective of the intended uses of the model. An object, process, or phenomenon that is simulated is referred to as simuland (Sokolowski and Banks, 2009). Cook and Skinner (2005) discuss different model validation methods and they include informal, static, dynamic, and formal. Informal methods are simple auditing, face validation (evaluation of the model by knowledgeable individuals), static methods check models without simulation or execution, dynamic methods evaluate the results of the model, and finally the formal methods rely on mathematical reasoning, inference, and proofs of correctness (Cook and Skinner, 2005).

Two important assumptions are also used while constructing the SD models, and they are: (a) processes are modeled to form a closed-loop (b) boundary of the system influences the dynamics. The models developed using the system dynamics are validated by several tests that focus on (1) replication; (2) sensitivity; and (3) prediction. These tests will confirm the structure of the model with the physical system that is being modeled. Various other tests that validate the dimensional consistency of the equations and robustness of the model in handling extreme conditions should be carried out before the model is adopted for implementation. A two-step process involving verification and validation can be carried out to confirm the robustness of the SD models. The models developed using the concepts of SD can be validated using all these tests. Bala et al. (2017) suggest several tests for structure, behavior, and policy implications and they are:

- Structure verification or confirmation tests can help check if the system replicates the real-life conditions correctly. These tests also help to check if the physics of the system is correctly represented in the model. These tests must confirm that the model structure does not contradict the existing knowledge about the real-world system (Duggan, 2016).
- Parameter verification tests can help check if the parameters of the model are chosen appropriately based on domain knowledge or based on a review of existing literature or real-life observations. The relationships, constants, and initial values used for stocks are also evaluated for correctness.
- Extreme condition tests can be used to evaluate the adequacy of the model in replicating extreme conditions that were observed in real life. Extreme conditions arising during the simulation can make functional relationships indeterminate and set variable values beyond their acceptable bounds. These tests will also help validate functional relationships.
- Boundary adequacy tests can help assess the correctness of system boundaries for simulating the process. Based on these tests, system boundaries can be expanded.
- A dimensional consistency test can be used to check the dimensional homogeneity of every equation (i.e., relationship) and units of the variables.
- Behavior reproduction, anomaly, and sensitivity tests are used to assess the model's capability to replicate the behavior of the system or how well the model's results match with those observed. Performance and error measures can be used to evaluate the model for its ability to characterize the real-life system behavior accurately. The anomaly test will be helpful to check if the assumptions used in the model development are correct, and the model-based anomalous behavior or response of the system is linked to wrong assumptions. Sensitivity tests are used to ascertain the model parameter uncertainty, and any

changes noted in the system performance for changes in parameter values are appropriate.

- Changed behavior prediction and policy-sensitivity tests evaluate the model prediction ability if a specific policy (i.e., the condition specified in the model) is changed and how sensitive the model response is to minor or major changes in the policy.
- Domain-specific knowledge tests take advantage of the expertise of professionals and stakeholders familiar with the system for validation of the SD models.

All these tests can be used during the model calibration and validation phase. Barlas (1996) has provided an extensive review of the SD model validation using many tests discussed earlier. In some fields, the validation process is also referred to as benchmarking, which refers to the exercise of comparing model results with the accepted representation of the process being modeled (DOD, 2011). A comprehensive review of general computer simulation verification and validation procedures can be found in works by Beisbart and Saam (2019) and Oberkampf and Roy (2010).

In the case of model calibration, both manual and automated (i.e., optimization based) procedures can be used to obtain the best values of the parameters used in the SD model. This process is referred to as calibration optimization by Dangerfield (2009, 2013). Parameter sensitivity analysis using Latin Hypercube Sampling (LHS) procedures can be used to obtain the best parameter values for the model. Ford and Flynn (2005) discuss the use of statistical screening of most influential inputs to SD models using the concepts of LHS. Validation of the model will require an assessment of several error and performance measures observed and estimated or predicted values from simulation for one or more variables. These are correctly referred to as time-domain measures (Murray-Smith, 2015) as errors are based on time-matched (i.e., chronologically paired), observed, and model-predicted variable values. Some of these error measures include mean error (ME), mean squared error (MSE), root mean squared error (RMSE), mean absolute error (MAE), and mean relative error (MRE). The measures ME, MSE, RMSE, and MAE are referred to as scale-dependent measures as they provide information in the units of the variable that is being evaluated. Measures such as MRE are scale-independent (non-dimensional) error measures. Performance measures include the Pearson correlation coefficient and index of agreement. The list of measures provided here is not exhaustive and it is recommended that the reader refers to literature related to verification and validation of computer simulation models. Error and performance measures documented in the literature are field-specific although most of them are universal.

Additional problem or system-specific measures can be developed to validate SD models. More discussion about SD model validation can be found in works by Sterman (1984) and Schwaninger and Grösser (2009). The sensitivity of the model to the time interval selected for simulation needs to be evaluated. It is beneficial to build SD models in phases and test them while aggregating additional components to the already developed models. The partial model testing approach, as recommended by Homer (1983), involves testing a portion of the model before the complete model is built.

A comprehensive checklist for SD model development was provided by Lai and Wahba (2001). They suggest checking of units, naming of variables, proper usage of constants, parameters, appropriate time step (i.e., the interval for simulation), the validity of stock–flow connections, the rigidity of rule-based or logic statements, and several others. Best practices for building SD models are discussed by Malczynski (2011). If the SD model is used for simulation of a variable over a long time, then the simulated time series should replicate the real-world variation. A multi-step validation procedure involving checks for trend, autocorrelation, amplitudes of extremes, the time between these extremes, and minimum and maximum values as recommended by Barlas (1996) can be used. Duggan (2016) recommends the use of mutation testing that is common in the software engineering field. The testing evaluates the model response to multiple changes (i.e., mutations) made to the model variables, constants, and arithmetic and logic operators. This testing is similar to uncertainty or sensitivity analysis used for a robust evaluation of models. Balci (1998) discussed over 75 verification, validation, and testing techniques for evaluation of models in different fields. Many of these techniques apply to SD model evaluation and validation.

The time series of the model generated variable and real-world observed values can be compared to assess and validate the SD models. Parametric and nonparametric statistical methods (Teegavarapu, 2018b) can be employed to facilitate comparisons. Parametric two-sample t-test and F-test can be used to assess statistically significant differences between the two-time series (i.e., model generated and observed). A nonparametric version of the t-test, the Mann–Whitney U test, can be used if the assumptions required for the t-test are not satisfied by the datasets. Distribution similarity between two-time series can be evaluated using unpaired two-sample Chi-square or Kolmogorov–Smirnov (KS) tests. Trends based on two-time series can be compared using an equal slopes test.

1.10 A SIMPLE SD MODEL

The flow diagram identifying stocks, flows, connectors, and convertors for the complete SD model is shown in Figure 1.3. Connectors transmit information between any two elements of the system (e.g., flows, stocks), and convertors can store relationships (e.g., tabular or graphical) and constants. While the functionalities of connectors and converters are the same in many modeling environments and discussed in the context of SD, they

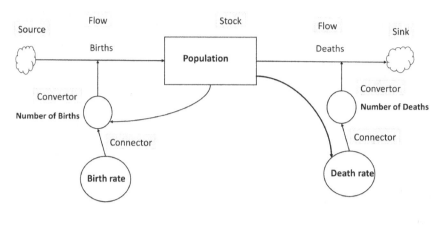

Flow Diagram

Figure 1.3 Flow diagram of the complete population model.

may be referred to with different names. The cloud objects in the flow diagram show the boundaries of the system identified by the source and sink. These objects can represent the origin and ending points of a flow, and in the simulation of some systems, the flows beyond these ending points are inconsequential to the system behavior beyond the boundary of the system.

Linear growth is caused by constant inflow (flow into the stock), and decay can be due to constant outflow from the stock. These behaviors can be generated using stock and flow. Exponential growth or decay can be simulated using a stock, flow, and convertor that revises flow based on the stock level or magnitude. S-shaped behavior can be simulated by using stock and flows (inflows and outflows) with inflows and outflows defined based on the stock level. Oscillatory behavior can be simulated using two stocks that influence each other through flows and connectors. The stocks are coupled and are dependent on each other. An excellent example of such behavior is known to exist in predator–prey systems.

1.11 GOVERNING EQUATIONS AND SOLUTION METHODS

The governing equations used for modeling different elements in a system are represented by finite difference expressions that are solved using standard numerical schemes. Euler and Runge–Kutta (RK) methods of integration are generally adopted for solving governing equations. The time interval for simulation is also an important aspect that will determine the accuracy of the numerical scheme used for solving the finite difference equations. The difference equations are solved using Euler or one of the two variants of RK methods (Press et al., 1992). The time interval of simulation (i.e., dt or Δt) needs to be specified to run the model and derive the changes in the system. The second- and fourth-order Runge–Kutta (RK4) methods are better than the Euler method for a given dt.

The stock value at any time can be estimated based on the initial value, flows (i.e., inflows and outflows), and the time interval Δt or dt. A simple mass balance equation can be written for stock using equation 1.6.

$$\text{Stock}_t = \text{Stock}_{t-1} + \int_{t-1}^{t} (\text{flow}_t^{in} - \text{flow}_t^{out}) \, dt \tag{1.6}$$

Equation 1.6 can be transformed into a differential form using equation 1.6.

$$\frac{d(\text{Stock})}{dt} = \text{flow}_t^{in} - \text{flow}_t^{out} \tag{1.7}$$

Equation 1.7 represents a standard mass balance equation that suggests that changes in stock (or stored value) over time dt is the difference in flows entering and leaving a system during the time interval. As the magnitude of the stock is required to evaluate the state of the system at any given time, the values of stock needed to be calculated. Equation 1.4 is solved as an initial value problem using Euler or Runge–Kutta method.

An illustration of function evaluation in the interval [0,4] using the Euler and fourth-order Runge–Kutta (RK4) method is provided in this section. The function is given by Equation 1.8.

$$z(t) = 3e^t \tag{1.8}$$

When $t_o = 0, z(0) = 3$ and the first partial derivative, $\frac{dz}{dt} = f(t,z) = z'(t) = 3e^t = z$.

Euler's method for iterative evaluation of function is given by Equation 1.9. The value of $f(t_o, z_o) = 3$ and for $n = 0,1...N$; the value of the function is calculated at every evaluation point, n. The variable Δt defines the increment used for the estimation of function values at regular intervals. This increment is the same as the time interval dt in Equation 1.10. The Euler method (also known as Euler forward method) for iterative of function is given by Equation 1.9.

$$z_{n+1} = z_n + \Delta t f(t_n, z_n) \quad \forall n \tag{1.9}$$

$$t_{n+1} = t_n + \Delta t \quad \forall n \tag{1.10}$$

The second-order Runge–Kutta (RK2) method for iterative evaluation of function is given by Equation 1.11.

$$z_{n+1} = z_n + \frac{1}{2}\Delta t\left(L_{n,1}^o + L_{n,2}^o\right) \quad \forall n \tag{1.11}$$

The values of $L_{n,1}^o$ and $L_{n,2}^o$ are calculated using Equations 1.12 and 1.13

$$L_{n,1}^o = f(t_n, z_n) \quad \forall n \tag{1.12}$$

$$L_{n,2}^o = f(t_n + \Delta t, z_n + L_1^o) \quad \forall n \tag{1.13}$$

The values of $L_{n,1}$, $L_{n,2}$, $L_{n,3}$, and $L_{n,4}$ are calculated using Equations 1.14–1.17.

$$L_{n,1} = f(t_n, z_n) \quad \forall n \tag{1.14}$$

$$L_{n,2} = f\left(t_n + \frac{\Delta t}{2}, z_n + \Delta t\frac{L_{n,1}}{2}\right) \quad \forall n \tag{1.15}$$

$$L_{n,3} = f\left(t_n + \frac{\Delta t}{2}, z_n + \Delta t\frac{L_{n,2}}{2}\right) \quad \forall n \tag{1.16}$$

$$L_{n,4} = f(t_n + \Delta t, z_n + \Delta tL_{n,3}) \quad \forall n \tag{1.17}$$

The fourth-order Runge–Kutta method for iterative evaluation of function is given by Equation 1.18.

$$z_{n+1} = z_n + \frac{1}{6}\Delta t\left(L_{n,1} + 2L_{n,2} + 2L_{n,3} + L_{n,4}\right) \quad \forall n \tag{1.18}$$

The two methods (viz., Euler and fourth-order Runge–Kutta methods) are applied with two different Δt values, and they are shown in Figures 1.4 and 1.5. Also, when Δt approaches close to zero, then the function value approximated by Euler's method approaches that of the analytical or exact solution.

In the first and second experiments with Δt values equal to 0.5 and 0.1, respectively, the number of function evaluations (N) is equal to 8 and 40. It is evident from both Figures 1.4 and 1.5 that the RK4 method is more accurate than Euler and visual inspection suggests that the estimates from RK4 overlap on the values of function values. As the Δt value decreases the function evaluations using Euler and RK4 methods become closer to actual function values, as evident from Figure 1.5. The truncation errors based on Euler and fourth-order Runge–Kutta methods are equal to Δt^2 and Δt^5. Also, the errors due to RK4 will be lower even for larger Δt values when compared to those from Euler's method with lower Δt values. The model

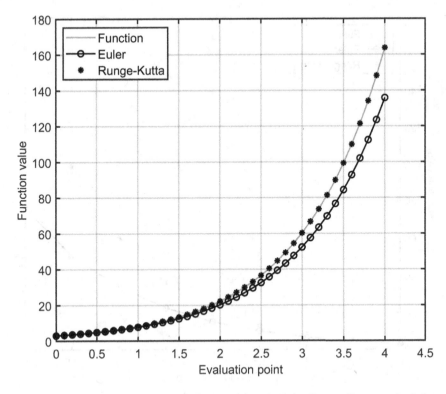

Figure 1.4 Function evaluations using Euler and fourth-order Runge–Kutta methods for Δt = 0.5.

results may be sensitive to the numerical integration method selected. Also, for a given Δt if RK2 or RK4 method produces similar results as those from Euler's method, then additional calculations required in RK2 and RK4 may not be beneficial from a computational point of view. The RK methods are ideal for the simulation of the systems in which variables change continuously and oscillatory behaviors are noted.

1.12 SD MODEL BUILDING ENVIRONMENTS

Several modeling commercial application software are available for building SD models and they are STELLA, Vensim, Powersim, and Simile (Muetzelfeldt and Massheder, 2003). Several recent books (i.e., Garcia, 2018a, 2020) explain the concepts of SD and provide practical exercises for building different models using commercially available modeling environments such as Vensim and STELLA. One of the oldest known modeling environments for system dynamics is DYNAMO (Richardson and Pugh III,

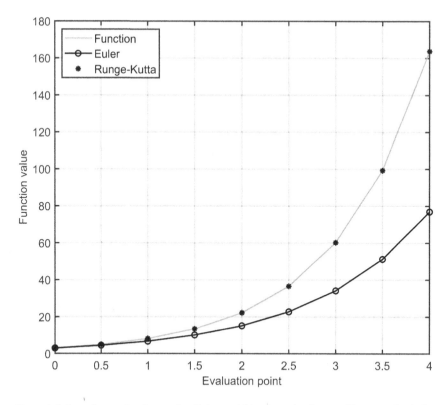

Figure 1.5 Function evaluations using Euler and fourth-order Runge–Kutta methods for Δ*t* = 0.1.

1981). The environment in which the system dynamics models are developed is referred to as an object-oriented simulation (OOS) environment. The reason being, the objects that are used possess generic properties and can be used to model elements with specific values. However, OOS should not be confused with object-oriented programming (OOP) nor OOS has the same abilities or features (e.g., inheritance, encapsulation, and polymorphism) as the OOP approach. An object-oriented simulation environment is a perfect tool to model any system with SD principles. Many of these environments have different layers with different features at which elements of the models can be viewed to look at the inner workings or numerical simulation aspects of the models. These environments provide an empty canvas referred to as modeling space. Drag-and-drop mechanisms or tools can help in using generic objects to build the SD model in the modeling space. Other modeling environments include AnyLogic, ithink from the makers of STELLA, DYNAMO, and Mapsim. A publicly available free online model development platform is Insight Maker (2020). Details about Insight Maker web-based modeling and simulation environment are provided by

Fortmann-Roe (2014). Basic modeling exercises using STELLA and Vensim are discussed by Martin (2001). Vensim and STELLA are popular environments for modeling water resources systems as reported in a recent survey by Mashaly and Fernald (2020). Vensim is most widely used SD development tool in water and environmental systems applications. Modelers have used Microsoft Excel and MATLAB for the development of SD models. Bates et al. (2019) use SD principles to build a model for groundwater storage model using Excel.

1.13 SD MODELING ENVIRONMENTS: STELLA AND INSIGHT MAKER

Two SD modeling environments (viz., STELLA and Insight Maker) are described in this section. The use of these environments in this book does not necessarily suggest an endorsement of any one of these two environments for the development of SD models. STELLA modeling environment provides three layers (viz., user interface, mapping, and modeling) of functionalities for users to develop models. The mapping layer provides the user interface, while the modeling layer is used for the construction of the model. Different objects are used in the modeling layer to develop the model. Three different numerical schemes (viz., Euler, second-order Runge–Kutta (RK2), and fourth-order Runge–Kutta (RK4)) with varying levels of accuracy are available. The equations are transparent to the user and can help understand various mathematical relationships. The governing equations are generic and are associated with the objects. For example, in the case of *stock* (object), a continuity equation for mass balance will be developed considering the inflows and outflows, while a *converter* carries a functional relationship between different variables that can be represented in a mathematical or a graphical form.

The simulation environment also provides several built-in mathematical, logical, and statistical functions that can be used in any of the objects. The governing equations based on the model structure are automatically created in the equation layer by the STELLA environment and can be reviewed for accuracy of the structure of the model. Functions such as random, if-then-else, and delay can be used to introduce unanticipated changes in the system and to construct rules/options. Table functions can be used if relationships between any two variables are too complex to represent in one single mathematical relationship. The environment also provides features such as sensitivity analysis, provision for graphical inputs, and a simulation model in which the inputs can be changed dynamically. The boundaries can be defined using *sources* and *sinks*. More details of recent versions of the STELLA environment are provided by the product website (www.isee-systems.com/store/products/feature-updates.aspx). STELLA environment is also available online for model development.

The Insight Maker (Insight Maker, 2020) is a web-based modeling environment that can be used for the development of SD models. It defines the objects with specific properties such as stocks, flows, variables, and links as primitives. The modeling environment provides several features and they include unit check mechanism, graphical inputs in converters, ghost object creation (a feature to help create a copy of an object or a primitive), vectorization capability that allows duplication of models to assess different conditions, and Monte Carlo (MC) simulation to aid sensitivity and uncertainty analysis. Insight Maker offers an optimization approach, a simple goal-seek mechanism, to obtain pre-specified targets in the model by changing the values of primitives. Insight Maker also provides users to build models using agent-based modeling (ABM) approaches and allows integration of ABM with SD models. The modeling environment also enables to develop the custom user interface and to hide and show different parts of the model in response to user actions (Insight Maker, 2020). It is also easy to share the model with others via the internet and work collaboratively with others. The modeling environment also provides several mathematical, time, random number functions.

A set of basic building blocks, "objects," used in the STELLA environment is shown in Figure 1.6. STELLA provides three layers (mapping, model building, and equation) that can be used to develop a complete model. These are linked to each other and any modification made in any one of the layers is reflected in all the other layers. The objects of Insight Maker are shown in Figure 1.7. An example of a stock and flow diagram for a water resources system (i.e., reservoir) is shown in Figure 1.8.

A model built using a web-based simulation model environment available in the public domain, Insight Maker (Insight Maker, 2020),

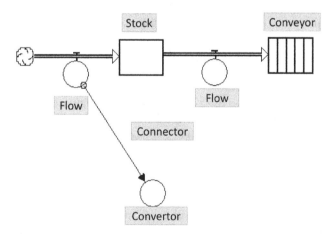

Figure 1.6 Main objects and functions in the STELLA simulation environment.

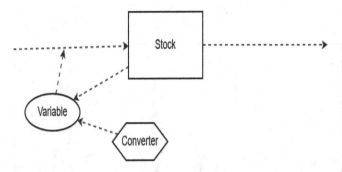

Figure 1.7 Building blocks (i.e., primitives) in the simulation environment, Insight Maker.

is shown in Figure 1.9. The stocks, flows, connectors (or links), and converters are shown. Computer simulations are finally performed using differential equations to integrate stocks and flows and to understand the behavior of the system. Results of the simulation model using fast Euler's method with an initial value of population as 10, birth rate as 0.6, and the death rate as a function of the population are shown in Figure 1.9. The Insight Maker environment also provides the fourth-order Runge–Kutta method (referred to as RK4) for the estimation of stock values at different time intervals. The population model developed in the STELLA environment is shown in Figure 1.10. It can be seen from Figures 1.9 and 1.10 that the population increases initially as exponential growth and then becomes stable represented by the "S" shaped curve. Loop dominance is evident as negative feedback dominates to control the population to a sustainable level.

1.13.1 Simulating Generic Behaviors

Generic behaviors of systems discussed in Section 1.8 are now discussed and demonstrated with stock and flow diagrams and simulations using SD models. These models are developed in the STELLA modeling environment and are inspired by generic behavior descriptions provided in many introductory SD books and work by Amadei (2015). The notation used for the equations is similar to the one used by Amadei (2015) as well. Nine different cases of system behavior are described, and simulation results are presented in this section. All the models are executed using a constant computational time interval except in the exercise that evaluates the accuracy of numerical methods.

Case 1: Linear Growth

This behavior is noted with constant inflow to a stock. Stock and flow diagram along with simulation results are shown in Figure 1.11. The initial

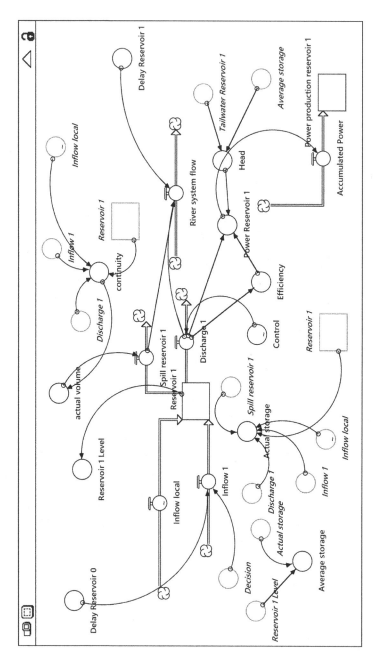

Figure 1.8 Reservoir operation problem represented by stocks and flows using an SD simulation environment, STELLA.

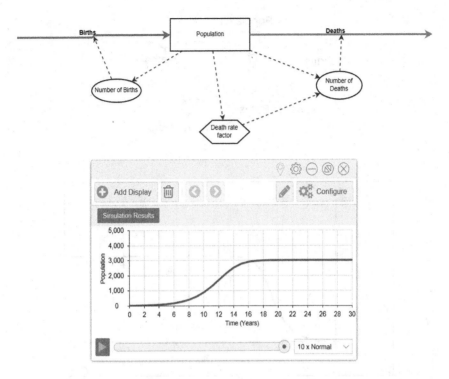

Figure 1.9 Population model built using the Insight Maker simulation environment along with the results of the simulation.

stock value (S_o) and constant inflow (C) will lead to linear stock growth with time (Δt), $S_t = S_o + C\Delta t$.

Case 2: Linear Decay

This behavior is noted with the constant outflow from the stock. Stock and flow diagram along with simulation results are shown in Figure 1.12. The initial stock value (S_o) and constant outflow (D) will lead to linear stock decay with time (Δt), $S_t = S_o - D\Delta t$.

Case 3: Linear Growth or Decay with Constant Inflows and Outflows

Linear growth and decay behaviors are noted in two different systems with constant inflows to and outflows from the stock. Stock and flow diagrams along with simulation results are shown in Figure 1.13. The initial stock value (S_o) and constant outflow (D) will lead to linear stock variation with time (Δt), $S_t = S_o + (C - D)\Delta t$. Linear growth results from inflow being

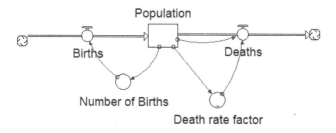

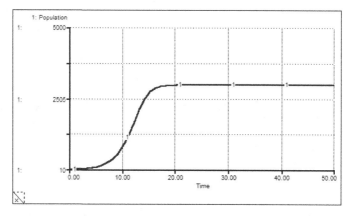

Figure 1.10 Population SD model built using the STELLA simulation environment along with the results of the simulation.

higher than outflow (i.e., $C > D$), and decay is noted when outflow is higher than inflow (i.e., $C < D$). No change in the stock from the initial value will be noted when inflow and outflow are equal (i.e., $C = D$).

Case 4: Exponential Growth

Exponential growth is noted in a system with inflow influenced by the level of the stock thus producing confounding behavior. Stock and flow diagrams along with simulation results are shown in Figure 1.14.

The sensitivity of computational time interval and numerical integration schemes for this system (i.e., exponent growth, with stock changing with time as $S(t) = S_o e^{k\Delta t}$) are evaluated. The solution schemes used in this exercise are Euler and two variants of Runge–Kutta (i.e., RK2 and RK4) methods with different intervals of simulation (i.e., Δt). The initial stock (S_o) and the growth factor (k) values used are 100 and 0.3, respectively. The results of the simulation are provided in Table 1.1. It can be observed even at low values of Δt; Euler's method performs inferior to Runge–Kutta methods.

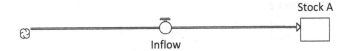

Stock A

Inflow

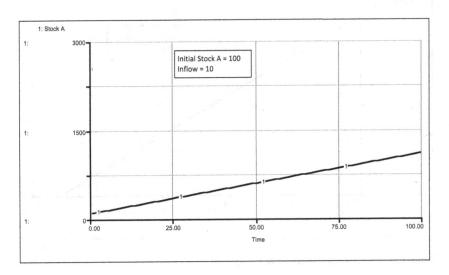

Figure 1.11 Linear growth behavior of the system.

Case 5: Exponential Growth with Varying
Inflows and Constant Outflows

Exponential growth is noted in a system with constant outflow and varying inflows influenced by the level of the stock thus producing confounding behavior. Stock and flow diagrams along with simulation results are shown in Figure 1.15.

Case 6: Exponential Decay

Exponential decay is noted in a system with outflow influenced by the level of the stock thus producing confounding behavior. Stock and flow diagrams along with simulation results are shown in Figure 1.16. The exponent growth is noted with stock changing with time as $S_t = S_o e^{-k\Delta t}$.

Case 7: Exponential Growth with
Varying Inflows and Outflows

Exponential growth is noted in a system with varying outflows and inflows influenced by the level of the stock thus producing confounding behavior similar

Stock B

Outflow

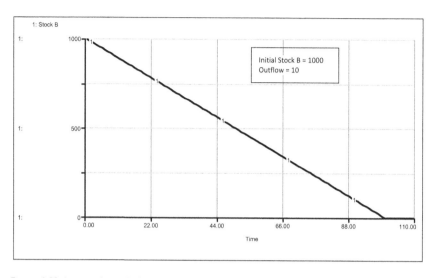

Figure 1.12 Linear decay behavior of the system.

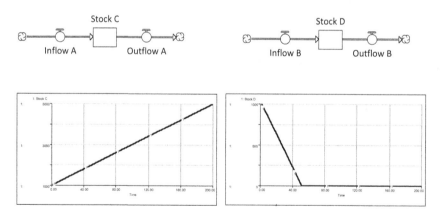

Stock C

Inflow A   Outflow A

Stock D

Inflow B   Outflow B

Figure 1.13 Linear growth and decay behaviors of two systems.

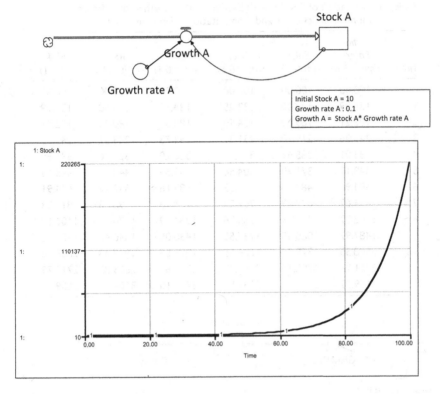

Figure 1.14 Exponential growth behavior of the system.

to Case 5. The growth of the stock now also affected by varying outflow. Stock and flow diagrams along with simulation results are shown in Figure 1.17.

Case 8: Goal-seeking Behavior

Goal-seeking behavior is noted in a system with constant inflow influenced by the difference in the level of the stock and the pre-specified target level required for the stock. The difference (or gap) is multiplied by a factor as an adjustment made to the stock level. An increase in this factor leads to increases in the adjustments made to the stock over time. The flow requires to be bi-directional as shown in Figure 1.18 to allow adjustments. Stock and flow diagrams along with simulation results are shown in Figure 1.18. The stock varies with respect to time (Δt) as $S_t = T + (S_o - T)e^{-ar\Delta t}$, with T as the target and ar as the adjustment rate.

Case 9: Logistic S-Shaped Growth Behavior

The growth behavior of stock when restriction exists due to capacity constraint is more often noted in studies involving growth and decline in

Table 1.1 Simulation of System with Exponential Growth with Different
           Numerical Schemes and Computational Time Intervals

Time	Function-Based Estimation	Euler (dt = 1)	Euler (dt = 0.5)	Euler (dt = 0.1)	RK2 (dt = 1)	RK4 (dt = 1)
0	100.00	100.00	100.00	100.00	100.00	100.00
1	134.99	130.00	132.25	134.39	134.50	134.98
2	182.21	169.00	174.90	180.61	180.90	182.21
3	245.96	219.70	231.31	242.73	243.31	245.95
4	332.01	285.61	305.90	326.20	327.26	331.99
5	448.17	371.29	404.56	438.39	440.16	448.13
6	604.96	482.68	535.03	589.16	592.02	604.91
7	816.62	627.49	707.57	791.78	796.26	816.53
8	1102.32	815.73	935.76	1064.09	1070.97	1102.18
9	1487.97	1060.45	1237.55	1430.05	1440.46	1487.76
10	2008.55	1378.58	1636.65	1921.86	1937.42	2008.24
11	2711.26	1792.16	2164.47	2582.82	2605.82	2710.79
12	3659.82	2329.81	2862.52	3471.10	3504.83	3659.13

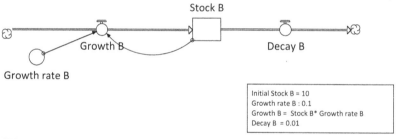

Initial Stock B = 10
Growth rate B : 0.1
Growth B = Stock B* Growth rate B
Decay B = 0.01

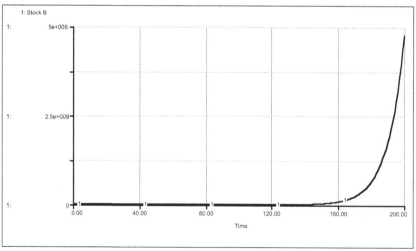

Figure 1.15 Exponential growth behavior of the system influenced by varying inflows and
             constant outflow.

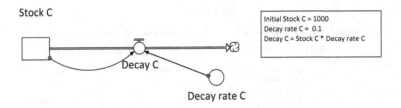

Stock C

Initial Stock C = 1000
Decay rate C = 0.1
Decay C = Stock C * Decay rate C

Decay C

Decay rate C

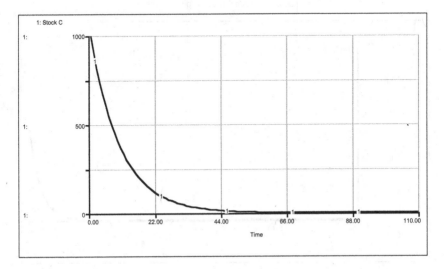

Figure 1.16 Exponential decay behavior of the system.

human and animal populations. This behavior is depicted using a stock and flow diagram. Logistic S-shaped (or Sigmoidal) growth is seen with the initial exponential growth of the stock. The stock varies with respect to time (Δt) as $S_t = AC / (1 + [AC / S_o - 1]e^{-ar\Delta t})$, with AC as the capacity and *ar* as the growth rate. The initial stock (S_o) value, growth rate (*ar*), and available capacity (AC) values are 500, 0.3, and 1000, respectively. The stock and flow diagram of the SD model and simulation results are presented in Figure 1.19. This model simulates only the growth side of the system (e.g., population model without deaths). However, the growth rate is controlled by the capacity-dependent constraint.

1.14 DIMENSIONAL CONSISTENCY AND SOLUTION TIME INTERVAL

Variables used in developing SD models have different units of measure, and therefore dimensional consistency (or unit consistency) needs to be checked to ensure that the equations or relationships used are dimensionally

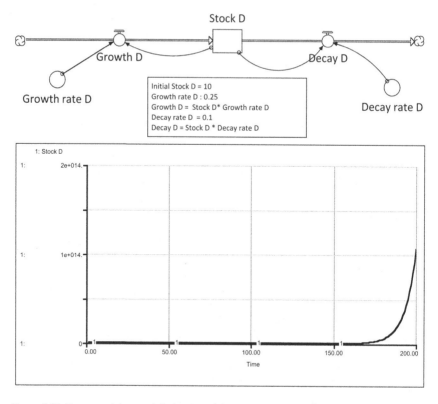

Figure 1.17 Exponential growth behavior of the system influenced by varying inflows and outflows.

homogeneous. For example, dimensional consistency is ensured when flows attached to stocks should have a unit that is expressed as a quantity that changes with time. Often converters are used to maintain the correct dimensionality. Units check is provided as one of the important features available in many SD modeling environments. Also, almost all SD modeling environments provide a facility to evaluate equations for dimensional constancy checks. Gary (1992) indicates that dimensionally impossible relationships should not be defined based on unclear mental models.

The response of the system when simulated and changes in the variable values are derived for a specific time interval (Δt) set by the modeler. This interval is also referred to as the computational interval, and in general, as Δt decreases, the integration error reduces and the number of numerical calculations increases. Many SD modeling environments allow users to modify the magnitude of this interval and select one of the numerical solution schemes that are available within the modeling environment. The accuracy of the numerical schemes used for the solution is sensitive to this time interval. This interval for simulation should be judiciously chosen

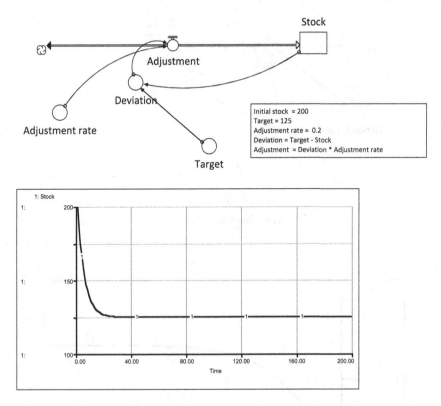

Figure 1.18 Goal-seeking behavior of the system influenced by inflow with a pre-specified target level (*T*) of the stock.

considering the type of numerical scheme to be used and time introduced by any delay(s) introduced in the system by one more variable. One recommendation is to use a value for a time interval (Δt) that is less than half of the shortest first-order delay in any system (Breierova, 1998). Many modeling environments suggest an appropriate time interval that is a compromise between accuracy and time taken by simulation (or speed of simulation).

1.15 MODELING ENVIRONMENTS AND ISSUES OF SD APPROACH

Modeling environments listed in the previous section provide several features that will help in the development of SD models. Commercially available SD modeling environments have improved by incorporating new tools and functionalities in the past decade. However, some of these environments lack built-in features to model systems in the spatial domain, advanced uncertainty analysis, and optimization capabilities (e.g., automatic

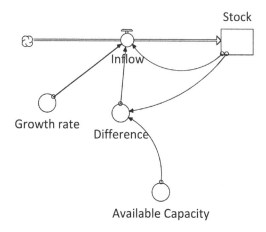

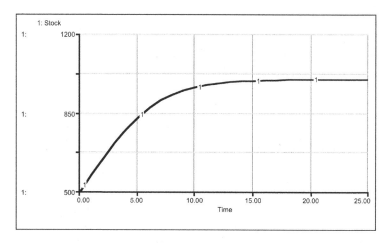

Figure 1.19 Logistical S-shaped growth behavior of the system influenced by inflow with pre-specified upper limit (available capacity) stock.

calibration facility) to obtain the best model parameters. Many of the validation checks discussed earlier in this chapter are not currently available as standard features in the current SD modeling environments. In many cases, the SD modeling environments are combined with other approaches or techniques to overcome some of the limitations listed here. SD models can be developed using different programming languages (e.g., R, Python) if the users are familiar with writing codes in those languages. Duggan (2016) demonstrates the use of the R program language to implement SD models. In some cases, these computer programming languages provide several advantages compared to restrictive drag-and-drop object-based modeling environments. The programming languages also provide a suite of numerical methods (e.g., Runge–Kutta) for the solution of difference equations within SD models. For example, the library package *deSolve* (Lin et al., 2020; Soetaert et al., 2010) written in R programming language can be used for the solution of differential equations.

A compilation of criticisms of the system dynamics approach and models was provided by Featherston and Doolan (2012) and Lin et al. (2020). Issues discussed were related to the role of the use of data for calibration, the inability of SD models to replicate reality, and predicted often falling into the trap of reductionist approaches which only deal only with the interaction between the elements of systems. The system dynamics approach mainly considers the endogenous perspective which considers that interactions among different elements of the system are responsible for system behavior. However, exogenous variables do exist, and they need to be considered and connected to internal elements of the system. Variables that are referred to as "coenetic" influence the closed-boundary system, but also other processes outside of the system do exist. These variables need to be recognized while building SD models and also expanding the boundaries of the system under consideration.

Adams and Hester (2012) document seven types of errors that are possible while adopting systems thinking–based approaches. These errors, according to Adams and Hester (2012) and Hester and Adams (2014), are (1) false positive (Type I); (2) false negative (Type II); (3) precise solution for a wrong problem (Type III); (4) wrong action (Type IV); (5) inaction (Type V); (6) questionable inference (Type VI); and (7) a combination of errors 1–6 (Type VII). Applications of the SD approach are not immune to these types of errors. Errors 1–3 occur during the model development phase and they can be corrected by a better understanding of the system and additional data collection and inclusion of variables that were not deemed essential initially. Errors 4–6 are mainly associated with the implementation of the policies, operational strategies, and decisions derived based on simulation model results. When SD models are used to understand and simulate systems, several problems, as identified by Funke (1991), will arise. These problems, according to Funke (1991), are described as (1) intransparency or lack of information about all the components or variables of the system; (2) polytely defined as a situation with multiple goals as provided by several stakeholders involved in the model building process; (3) complexity related to the number of variables and relationships among the variables; (4) variable influences as characterized by one-to-many or many-to-one relationships; (5) evolving systems that change with time and need to address the problems within specific time constraints; and (6) delays associated with time or temporal lags. Systems can be simulated using SD models considering enormous detail and dynamic complexities. However, considering the law or principle of parsimony (referred to as "Occam's or Ockham's razor") (Sober, 2015), models should be built with as few parameters and assumptions as possible with a focus on causation than on correlation. SD approach does not easily allow for seamless integration with other methodologies. Its high abstraction level dealing with aggregates, loops, and minimal details at macro-level and endogenous nature of variables and their influences limits its use (Borshchev, 2013).

1.16 SD APPROACH AND MODELING WATER AND ENVIRONMENTAL SYSTEMS

Water resources and hydrologic systems are governed by processes that can be modeled as interlinked components that store and transport water in space and time. Water resides in a state or a component of the hydrologic system for a specific amount of time (referred to as residence time) and this time varies from seconds to hundreds of years. Mass balance (change in storage (ΔS) = inflows (I) – outflows (O)) is always satisfied in space and time and becomes the quintessential relationship between flows and stored water at any point (both space and time) in the hydrosystem. The SD approach with flows and stocks as main building blocks is ideally suited for modeling water, hydrologic, and environmental systems. SD models for water resources modeling applications are mainly useful for understanding the basic behavior and scenario-based simulation.

The SD approach is also suitable for modeling environmental processes at different spatial and temporal scales. It is important to evaluate the appropriateness of an SD approach for characterizing the behavior of any system considering the availability of models that simulate the changes in different elements of the system at a micro-level as opposed to the macro-level. SD models are more appropriate for the latter.

Generic behaviors discussed earlier in this chapter are noted when water resources and environmental systems are evaluated through simulation. Water resources system infrastructure is generally operated and managed for the best possible spatial and temporal distribution of water considering the variability of demands and available water in space and time. Inputs to water and environmental systems are random and the relationship between any two variables may also change over time. The inputs include a time series of precipitation and streamflow values with or without strong autocorrelation. Strong and weak autocorrelation is noted for daily streamflows and precipitation, respectively. Streamflow series at monthly scale exhibit seasonality with an oscillatory pattern. Constant streamflow values are realized due to the existence of hydrologic/hydraulic structures controlling the flows. Examples of generic behaviors include nonlinear growth in the reservoir level (i.e., storage or height of the water) due to rapid influx of water into the reservoir due to an extreme precipitation event and goal-seeking behavior induced by human-controlled releases to maintain reasonable safe reservoir storage (or level) after extreme inflows. Overshoot and collapse can be seen in open channel cross-sections that provide temporary storage (i.e., wedge storage) for water during two stages (rising and falling water levels) of flooding events. Sustained oscillatory or cyclical behavior can be seen based on the operations of reservoirs to control releases based on seasonal water demands. The releases are subject to water availability in the reservoir. Growth and overshoot behavior can be evident in a storage of a reservoir with a spillway due to variations in the inflows that result in storage that exceeds the maximum capacity of the reservoir. The sigmoidal growth behavior that is common in natural systems is also possible in reservoir systems with capacity constraint (i.e., maximum available storage) and human interventions that introduce balancing feedback (or negative feedback) via enforcement of operational rules. Some of these behaviors are also noted in environmental systems. Diurnal variations of dissolved oxygen (DO) levels at different temporal scales in water bodies is an example of oscillatory behavior. Exponential growth in pollutant loads due to point and nonpoint sources in a watershed and reduction of these loads due to capacity restrictions of water bodies or management measures is a case of S-shaped growth.

Natural as well as engineered or managed water and environmental systems can be modeled using the SD approach. The SD approach using GUI-enabled software provides the most non-restrictive environment for modeling different processes of water and environmental systems. Unlike simulation models developed using high-level computing languages or spreadsheet-based environments (Guerrero, 2019), SD-based models provide the user with a visual conceptualization of different components of the systems. Interactions between storage and flow elements that form the crux of any water resources system are easy to understand and transparent to the user. SD models have the self-documenting feature which makes the variables and connections self-explanatory. The difference equations generated based on stock and flow diagrams are also available to the users in SD models for additional scrutiny. Water and environmental system-specific modeling objects that are part of drag-and-drop simulation environments (e.g., HEC-HMS) lack the flexibility to model any process that affects these systems. Due to the specific nature or properties of these objects, restrictions are also placed by modeling environments that dictate how objects can be connected. Unlike SD-supported modeling environments, many water resources simulation environments do not provide features to build graphical user interfaces (GUIs) to facilitate communication of simulation results to decision-makers. Modeling of engineered or human-controlled water and environmental systems using SD will require operation rules that are defined using "if-then-else" constructs. It can be argued that the inclusion of these constructs may lead to rigidity in modeling due to discontinuity in variable value assignments. Managed complex systems can also lead to unintended consequences due to human-controlled operations. Examples of these consequences are detrimental effects of construction of a reservoir (dam) on a river caused by (1) increased impairment of stream because of pollutants due to managed flows downstream of a reservoir; (2) inundation and ecological damage upstream of the reservoir; (3) changes to water availability to aquatic habitat; and (4) alternation of natural stream conditions and several others related to water loss and increased risk of catastrophic flooding due to dam failure. These consequences are often not assessed during the planning stage or sometimes neglected as the benefits of storing water in a reservoir outweigh the unintended immediate or long-term detrimental outcomes. Unintended negative consequences can be reduced if not eliminated by having knowledge about the system and the results from the iterative assessment of system behavior under different scenarios. Adaptive operational policies can be developed to manage complex systems in the short and long term.

SD models can also incorporate soft variables (e.g., preferences of decision-makers or operators of water resources systems) to enable the development of policies for improving system performance. Water and environmental systems are intricately linked with socio-economic systems along with biological, geophysical, and economical systems. Successful applications of SD-based approaches for socio-economical systems point to major benefits of modeling water and environmental systems using SD principles. The SD-based approach complements existing modeling and simulation methodologies to answer policy questions (Lin et al., 2020) which are critical for water resources management. Both forecasting and backcasting of system behavior are essential for the planning and operation of water resources systems and the SD approach can provide tools for these two tasks. SD simulation models can be developed for single and continuous events that are commonly considered in the hydrologic design and modeling efforts. Simulation studies involving SD models can benefit from a seven-step approach developed by Law (2005) for a successful simulation exercise. These steps include formulating the problem, collecting available data and constructing assumptions document, validating assumptions document, developing and validating the simulation model, designing, conducting, and analyzing different experiments, and finally documenting and presenting the simulation results.

While the SD approach is ideally suitable for developing simulation models for water and environmental systems, it is important to note the limitations of SD modeling environments. SD approach is not a panacea for all simulation modeling needs targeted to understand these systems. However, the SD approach offers all the required basic tools for building models for water and environmental systems as documented by a comprehensive review by Mashaly and Fernald (2020). There is difficulty in moving from an aggregate behavior mode analysis usually handled by the SD approach to disaggregate to individual-based modeling that is better handled by agent-based approaches. However, a few modeling environments built for SD simulation provide disaggregation tools (e.g., model within model or sub-model conceptualization). Water and environmental systems are heavily linked with and influenced by sociological, ecological, environmental, and political systems. The SD approach would be beneficial in integrating these systems with water and environmental systems. In some instances, a combination of multiple methods is required to achieve such integration. As the SD approach focuses mainly on the overall system behavior, individual (e.g., person, animal, etc.) behaviors or interactions between individuals can be addressed in the multimethod approach. Examples of multimethod approaches that combine system dynamics, agent-based modeling, and discrete-event simulation are provided by Borshchev (2013). The generic behavior patterns discussed in this chapter may not be always realized in water and environmental systems without socio-economic and other sector influences. Chapters 2 and 3 will provide an extensive review of the development of models for water and environmental systems and applications of the SD approach, respectively.

I.17 SD FOR LEARNER-CENTERED EDUCATION AND RESEARCH

Ever since its inception, the SD approach has been used to develop learner-centered education methods or simulation tools for improving the understanding of complex and dynamic systems. Learner-centered education promotes a unique environment in which a learner (i.e., student) works with a teacher or instructor to acquire knowledge, with the latter serving more as a facilitator than as an instructor. In such a learning environment, students can create tools that facilitate personal experimentation of the system behavior (Bowen et al., 2014; Bayrakceken and Arisoy, 2013; Wolstenholme and Coyle, 1983). The System Dynamics in Education Project (SDEP) (http://web.mit.edu/sysdyn/sdep.html), initiated and developed in 1980 at Massachusetts Institute of Technology (MIT) under the supervision of J. W. Forrester, has contributed and promoted several education initiatives and developed content for the promotion and use of systems thinking concepts in K–12 education. The website http://web.mit.edu/sysdyn/road-maps/toc .html provides a comprehensive description of the SD approach and model development. SD approach helps in the development of double-loop learning (DLL) in which the information feedback helps to continuously improve the model structure itself. DLL is opposite to single-loop learning (SLL) which relies on the constant or static model structure.

SD approach promotes learner-centered as opposed to teacher-centered education, and it facilitates participatory learning. Several simulation games using the SD approach were developed that promoted understanding of complex systems and allowed users to evaluate different modes of system behavior. SD supports the concept of modeling for learning by providing tools for simulation gaming (management flight simulators) (Barlas, 2002).

A management flight simulator serves as a visual front end for an SD model with a graphical user interface (GUI) that allows decision-makers to modify the parameters of the model, assess the evolution of the system behavior over time, and change the course of the system at any point of time. These models involve human interaction or require human in the loop (HITL) for simulations. Some of these management flight simulators allow multiple decision-makers to work on simulation thus allowing group-based decision-making. Examples of games that promote learning of complex systems with delays and feedback include the *"Beer Game"* and *Fish Banks Game*. The *Beer Game* (www.systemdynamics.org/beer-game) helps in understanding complex decision-making processes in the production/distribution system and the *Fish Banks Game* (Meadows et al., 2020) (www.systemdynamics .org/fish-banks-game; https://mitsloan.mit.edu/LearningEdge/simulations/ fishbanks/Pages/fish-banks.aspx) highlights the overexploitation of a common resource by multiple individuals with independent selfish goals of benefiting from the common resource. The *Beer Game* works as a replication of a real organization dealing with production and distribution systems whereas *Fish Banks Game* is similar to a flight simulator game. The system behavior in such a situation results in S-shaped growth, and the whole process ultimately results in a systems archetype referred to as "Tragedy of Commons" (Hardin, 1968). An interesting example of simulating Hamlet in the classroom was documented by Lee (1992). Other models that deal with interactions between population, economy, ecology, and others include *World1* (Forrester, 1971b), *World2*, and *World3* (Meadows et al., 1992), which are large-scale SD-based models. Simonovic (2002) has developed the *WorldWater* model based on the *World3* model, which integrates water quality and quantity sectors with five others such as industrial growth, population, agriculture, economy nonrenewable resources, and pollution. The readers are referred to a comprehensive review of the *World1* model provided by Jantsch (1971). An update on the *World3* model by Meadows et al. (2006) provided major insights gained over the three decades since *World1* was developed. Educational initiatives using the SD approach can be found in the works of Teegavarapu (2018a), Bier (2010), and Ford and Paynting (1995). The SD concepts can also substantially contribute to the evidence-based and knowledge-generating approaches to benefit the effectiveness of pedagogical techniques.

1.18 COMMUNITY-BASED SYSTEM DYNAMICS MODEL BUILDING

Model building for systems that are influenced, operated, and managed by several decision-makers can be a complex task. This task often requires involving all the stakeholders in the modeling development process and multiple phases of the model building, evaluation, and validation and improvement after implementation. This process is referred to as participatory system modeling or group model building that is appropriate for complex systems. This exercise allows participants to discuss their mental models and modify them according to the consensus of the group. Excellent discussions about the community-based SD model building process, engagement of communities, and building of teams to design, solve, and manage systems are provided by Hovmand (2014), Vennix (1999), and Richardson and Andersen (2010). Amadei (2015) describes the use of systems approach and SD models for developing community development projects. Belt and Dietz (2004) refers to the development of SD models for environmental consensus building as mediated modeling.

Community-based model building approach allows: (1) participatory model development; (2) consideration of more than one objective; (3) adaptive evolution and modification of the models; and (4) policy analysis and implementation strategy development considering aspiration levels of all the participants involving in model development and utilization. Setting up aspirations for an improved model development process involving a group may or may not lead to the so-called Pygmalion effect that relates improved performance to high expectations. SD models can be used interactively (or in flight-simulation mode) for analysis of different policies designed by modelers for managing the systems and evaluate if there is any resistance evident to these policies from the system behavior noted over time. Examples of participatory model development and simulation of complex water resources systems can be found in several studies (Vamvakeridou-Lyroudia and Savic, 2008; Stave, 2003; Beall et al., 2011; Videira et al., 2009). Community-based SD modeling efforts may not always lead to the best implementations of policies derived from the models as sometimes processes influencing the systems are not clearly understood. Such failures of policies can be attributed to Hanlon's razor principle, which refers to bad policy decisions made based on ignorance. Complementarity is another issue that needs to be considered when a group-based model building exercise is carried out as an understanding and modeling of a system may lead to agreements or disagreements among the model builders.

1.19 SUMMARY AND CONCLUSIONS

System behavior can be easily understood by developing descriptive models that can provide a transparent depiction of the system as well as all the elements that are part of it. Moreover, the interactions between different elements of the system via feedbacks that transfer both information as well as quantitative states of the different variables need to be clearly understood. System behavior in space and time needs to be modeled for real-time, short-, and long-term performance evaluation of the system along with a clear understanding of the dynamic nature of the behavior governed by the numerous feedbacks.

The SD approach discussed in this chapter provides generic building blocks to develop models to create, evaluate, and forecast the behavior of the systems over time. In many instances, SD models are mainly used for evaluating qualitative changes in the reference modes or expected modes of the system behavior. Counter-intuitive results and surprise behaviors in systems from SD model simulation results provide the first level of scrutiny to understand the system better and evolve better management policies. SD approach considers the behavior of the system at the aggregate level and not from an individual element point of view. Agent-based modeling approaches can be used to address the micro-level interactions between different elements of the system. SD approach promotes holism as opposed to reductionism in some way when systems are evaluated. This chapter provided a summary of the systems thinking, behavior, and fundamentals of the SD approach. The motivation for the use of the SD approach for modeling water and environmental systems is briefly elaborated. The next two chapters will discuss the process of building models that can help to understand the dynamic nature of water resources and environmental systems.

REFERENCES

Adams, K. M., P. T. Hester (2012). Errors in systems approaches. *Internal Journal of Systems Engineering*, 3(3/4), 233–242.

Adams, K. M., J. H. Mun (2005). The application of systems thinking and systems theory to systems engineering. *Proceedings of the 26th National ASEM Conference: Organizational Transformation: Opportunities and Challenges*, 493–500, American Society for Engineering Management, Virginia Beach.

Amadei, B. (2015). *A Systems Approach to Modeling Community Development Projects*. Momentum Press, New York.

Åström, K. J., R. M. Murray (2008). *Feedback Systems: An Introduction for Scientists and Engineers*. Princeton University Press, Princeton, NJ.

Bacaër, N. (2011). Verhulst and the logistic equation (1838). *A Short History of Mathematical Population Dynamics*, 35–39. Springer, London.

Bala, B. K. (1999). *Principles of System Dynamics*, First edition. Agrotech Publishing Academy, Udaipur.

Bala, B. K., F. M. Arshad, K. M. Noh (2017). *System Dynamics Modelling and Simulation*. Springer, Singapore.

Balci, O. (1998). Verification, validation, and testing. *The Handbook of Simulation*. J. Banks, Chapter 10, 335–393, John Wiley, New Jersey.

Barlas, Y. (1996). Formal aspects of model validity and validation in system dynamics. *System Dynamics Review*, 12(3), 183–210.

Barlas, Y. (2002). System dynamics: Systemic feedback modeling for policy analysis. *Knowledge for Sustainable Development, an Insight into the Encyclopedia of Life Support Systems*, 1, 1131–1175, UNESCO-EOLSS, Oxford.

Bates, G., M. Beruvides, C. Fedler (2019). System dynamics approach to groundwater storage modeling for basin-scale planning. *Water*, 11(9), 1–17. doi:10.3390/w11091907

Bayrakceken, M. K., A. Arisoy (2013). An educational setup for nonlinear control systems: Enhancing the motivation and learning in a targeted curriculum by experimental practices. *Control Systems*, IEEE, 33, 64–81.

Beall, A., F. Fiedler, J. Boll, B. Cosens (2011). Sustainable water resource management and participatory system dynamics. Case study: Developing the Palouse basin participatory model. *Sustainability*, 3, 720–742.

Bellinger, G. (2020). http://www.systems-thinking.org/arch/arch.htm (accessed June 2020).

Belt, D., T. Dietz (2004). *Mediated Modeling: A System Dynamics Approach to Environmental Consensus Building*. Belt, Island Press, Washington D.C.

Bier, A. (2010). Simulating a thermal water quality trading market for education and model development. *Journal of Environmental Management*, 91(12), 2491–2498.

Biesbart, C., N. J. Saam (2019). *Computer Simulation Validation: Fundamental Concepts, Methodological Frameworks, and Philosophical Perspectives*. Springer, Switzerland.

Birta, L. G., G. Arbez (2019). *Modelling and Simulation: Exploring Dynamic System Behaviour*. Springer, Switzerland.

Borshchev, A. (2013). *The Big Book of Simulation Modeling: Multimethod Modeling with AnyLogic 6*. AnyLogic, North America.

Borshchev, A., A. Filippov (2004). From system dynamics and discrete event to practical agent-based modeling: Reasons, techniques, tools. The 22nd International Conference *of the System Dynamics Society*, July 25–29, Oxford, England.

Bowen, J. D., D. N. Perry, C. D. Bell (2014). Hydrologic and water quality model development using Simulink, *Journal of Marine Science and Engineering*, 2(4), 616–632.

Brailsford, S., L. Churilov, B. Dangerfield (2014). *Discrete-Event Simulation and System Dynamics for Management Decision Making*. Wiley and Sons, Ltd., Chichester.

Breierova, L. (1998). Mistakes and misunderstandings: DT error, Report D-4695. MIT System Dynamics in Education Project. https://ocw.mit.edu/courses/sloan-school-of-management/15-988-system-dynamics-self-study-fall-1998-spring-1999/readings/mistakes6.pdf (accessed May 2020).

Bureš, V., T. Otčenášková, M. Zanker (2019). Mistakes in system dynamics models: An educational issue from the systems engineering perspective. *Proceedings of the International Conference on Industrial Engineering and Operations Management*, July 23–26, Pilsen, Czech Republic

Campuzano, F., J. Mula (2011). *Supply Chain Simulation: A System Dynamics Approach for Improving Performance*. Springer-Verlag, London.

Cavana, R. Y., E. D. Mares (2004). Integrating critical thinking and systems thinking: From premises to causal loops. *System Dynamics Review*, 20(3), 223–235.

Cook, D., J. M. Skinner (2005). How to perform credible verification, validation, and accreditation for modeling and simulation. *Systems and Software Technology Conference*, 20–24.

Coyle, R. G. (1996). *System Dynamics Modeling: A Practical Approach*. Springer Science, Germany.

Dangerfield, B. C. (2009). Optimization of system dynamics models. *Encyclopedia of Complexity and Systems Science*. R. A. Meyers, 9034–9043, Springer, New York.

Dangerfield, B. C. (2013). Systems thinking and system dynamics: A primer. *Discrete-Event Simulation and System Dynamics for Management Decision Making*, S. C. Brailsford, L. Chudinov, B. C. Dangerfield, 26–51. John Wiley & Sons Ltd., Chichester, U. K.

Deaton, M. L., J. J. Winebrake (2000). *Dynamic Modeling of Environmental Systems*. Springer, New York.

DOD (2011). Department of Defense. Modeling and Simulation (M&S) Glossary. https://www.acqnotes.com/Attachments/DoD%20M&S%20Glossary%201%20Oct%2011.pdf (accessed October 2020).

Doyle, J., D. Ford (1998). Mental Model Concepts for System Dynamics Research. *System Dynamics Review*, 14(1), 3–29.

Duggan, J. (2016). *System Dynamics Modeling with* R. Springer International, Switzerland.

Eisner, H. (2019). *Thinking: A Guide to Systems Engineering Problem-Solving*. CRC Press, Boca Raton, FL.

Esfandiari, R. S., B. Lu (2018). *Modeling and Analysis of Dynamic Systems*. CRC Press, Boca Raton, FL.

Featherston, C., M. Doolan (2012). A critical review of the criticisms of system dynamics. The 30th International Conference *of the System Dynamics Society*, St. Gallen, Switzerland, July 22. https://openresearch-repository.anu.edu.au/bitstream/1885/18409/4/02_Featherston_A_Critical_Review:of_the_2012.pdf (accessed May 2020).

Fieguth, P. (2017). *An Introduction to Complex Systems: Society, Ecology and Nonlinear Dynamics*. Springer, Switzerland.

Ford, A. (1999). *Modeling the Environment: An Introduction to System Dynamics: An Introduction to System Dynamics Models of Environmental Systems*, Island Press, Washington, DC.

Ford, A., H. Flynn (2005). Statistical screening of system dynamics models. *System Dynamics Review*, 21(4), 273–302.

Ford, D. N. (2018). A system dynamics glossary. https://sds.memberclicks.net/assets/SDGlossary.pdf (accessed May 2020).

Ford, D. N., R. Paynting (1995). Linking academic theory and industry practice with student interactive projects. *The Center for Quality Management Journal*, 4(2), 19–32.

Forrester, J. W. (1958). Industrial dynamics: A major breakthrough for decision makers. *Harvard Business Review*, 36(4), 37–66.

Forrester, J. W. (1961). *Industrial Dynamics*. The MIT Press, Cambridge, MA.

Forrester, J. W. (1968). *Principles of Systems*. Productivity Press, Portland, OR.

Forrester, J. W. (1971a). *World Dynamics*. Pegasus Communications, Waltham, MA.

Forrester, J. W. (1971b). Counterintuitive behavior of social systems. *MIT Technology Review*, 73(3), 52–68.

Forrester, J. W. (1994). Systems thinking and soft OR. *System Dynamics Review*, 10(2), 245–256.

Forrester, J. W. (1995). Counterintuitive behavior of social systems. https://ocw.mit.edu/courses/sloan-school-of-management/15-988-system-dynamics-self-study-fall-1998-spring-1999/readings/behavior.pdf

Forrester, J. W. (2009). *Some Basic Concepts in System Dynamics*. notes, Sloan School of Management, Massachusetts Institute of Technology (MIT).

Fortmann-Roe, S. (2014). Insight maker: A general-purpose tool for web-based modeling & Simulation. *Simulation Modelling Practice and Theory*, 47, 28–45.

Funke, J. (1991). Solving complex problems: Exploration and control of complex systems. *Complex Problem Solving: Principles and Mechanisms*, 185–222, R. J. Sternberg, P. A. Frensch, Lawrence Erlbaum Associates, Hillsdale, NJ.

Garcia, J. M. (2018a). *System Dynamic Modelling with Vensim: A Book for Learning the Applications of Simulation Models to Manage Complex Feedback Control*. Innova Books, Spain.

Garcia, J. M. (2018b). *Feedbacks. From Causal Diagrams to System Thinking: Manage Dynamical Systems in Business, Ecology, Biology and Social Sciences, Using Balancing and Reinforcing Loops*. Innova Books, Spain.

Garcia, J. M. (2019). *Modeling the Economy: Money and Finances: Selected Papers on System Dynamics. A Book Written by Experts for Beginners (Analysis and Optimization)*. Independently published.

Garcia, J. M. (2020). *Common Mistakes in System Dynamics: Manual to Create Simulation Models for Business Dynamics, Environment and Social Sciences (Vensim)*. Independently Published.

Gary, M. S. (1992). Mistakes and misunderstandings: Examining dimensional inconsistency. D-4452-1. MIT system dynamics in education project. https://ocw.mit.edu/courses/sloan-school-of-management/15-988-system-dynamics-self-study-fall-1998-spring-999/readings/mistakes.pdf (accessed May 2020).

Ghosh, A. (2017). *Dynamic Systems for Everyone: Understanding How Our World Works*. Springer, Switzerland.

Goodman, M. R. (1989). *Study Notes in System Dynamics*. Pegasus Communications, Waltham, MA.

Gordon, S. I., B. Guilfoos (2017). *Introduction to Modeling and Simulation with MATLAB and Python*. CRC Press, Boca Raton, FL.

Guerrero, H. (2019). *Excel Data Analysis: Modeling and Simulation*. Springer Nature, Switzerland.

Hardin, G. (1968). The tragedy of the commons. *Science* 162(3859), 1243–1248.

Hester, P. T., K. M. Adams (2014). *Systemic Thinking: Fundamentals for Understanding Problems and Messes*. Springer, Switzerland.

Homer, J. B. (1983). Partial-model testing as a validation tool for system dynamics. *International System Dynamics Conference Proceedings*, 920–932, Chestnut Hill, MA.

Hovmand, P. S. (2014). *Community-Based System Dynamics*. Springer-Verlag, New York.

Insight Maker (2020). https://insightmaker.com/ (accessed May 20, 2020).

Jantsch, E. (1971). World dynamics. *Futures*, 3(2), 162–169.

Kauffman, D. L (1980). *Systems One: An Introduction to Systems Thinking (Identifying the Larger Pattern of Interconnections)*. Innovative Learning Series, Future Systems, Inc.

Keating, E. K. (2020). Everything you ever wanted to know about how to develop a system dynamics model but were afraid to ask. https://proceedings.systemdynamics.org/1998/PROCEED/00024.PDF (accessed October 2020)

Kim, D. H. (1992). Toolbox: Guidelines for drawing causal loop diagrams. *The Systems Thinker*, 3(1), 5–6.

Kim, D. H. (2000). *Systems Thinking Tools: A User's Reference Guide*. Pegasus Communications, Inc., Massachusetts.

Kirkwood, C. W. (1998). System behavior and causal loop diagrams. *System Dynamics Methods: A Quick Introduction*, in System Dynamics Methods: A Quick Introduction, 1–14. https://www.public.asu.edu/~kirkwood/sysdyn/SDIntro/ch-1.pdf (accessed October 2020).

Korzybski, A. (1994). *Science and Sanity: An Introduction to Non-Aristotelian Systems and General Semantics*. 5th ed. Institute of General Semantics, Englewood.

Kunc, M. (2018). *System Dynamics: Soft and Hard Operational Research*. Palgrave Macmillan, London.

Lai, D., R. Wahba (2001). System dynamics model correctness checklist. https://ocw.mit.edu/courses/sloan-school-of-management/15-988-system-dynamics-self-study-fall-1998-spring-1999/readings/checklist.pdf (accessed May 2020).

Law, A. M. (2005). How to build valid and credible simulation models. *Proceedings of the Winter Simulation Conference*, Orlando, FL, USA, 2005, IEEE. 1–9, doi: 10.1109/WSC.2005.1574236.

Lee, H. P. (1992). Simulating hamlet in the classroom. *System Dynamics Review*, 8(1), 91–98.

Lin, G., M. Palopoli, V. Dadwal (2020). From causal loop diagrams to system dynamics models in a data-rich ecosystem. *Leveraging Data Science for Global Health*, L. Celi, M. Majumder, P. Ordóñez, J. Osorio, K. Paik, M. Somai, Springer, Cham. https://doi.org/10.1007/978-3-030-47994-7_6 (accessed May 2020).

Maani, K. E, R. Y. Cavana (2000). *Systems Thinking and Modelling: Understanding Change and Complexity*. Prentice-Hall, Auckland.

Malczynski, L. A. (2011). Best practices for system dynamics model design and construction with Powersim Studio. Sandia National Laboratories Report. https://prod-ng.sandia.gov/techlib-noauth/access-control.cgi/2011/114108.pdf (accessed May 2020).

Martin, L. A. (1997). An introduction to feedback. https://ocw.mit.edu/courses/sloan-school-of-management/15-988-system-dynamics-self-study-fall-1998-spring-1999/readings/feedback.pdf (accessed May 2020).

Martin, L. A. (2001). Beginner modeling exercises, https://ocw.mit.edu/courses/sloan-school-of-management/15-988-system-dynamics-self-study-fall-1998-spring-1999/readings/modeling.pdf (accessed May 2020).

Mashaly, A. F., A. G. Fernald (2020). Identifying capabilities and potentials of system dynamics in hydrology and water resources as a promising modeling approach for water management. *Water*, 12, 1432. doi:10.3390/w12051432.

Masuch, M. (1985). Vicious circles in organizations. *Administrative Science Quarterly*, 30(1), 14–33.

Masys, A. J. (2016). *Applications of Systems Thinking and Soft Operations Research in Managing Complexity, from Problem Framing to Problem Solving*. Springer, Switzerland.

Meadows, D. H. (2008). *Thinking in Systems: A Primer*. Earthscan, New York.

Meadows, D. H., D. L. Meadows, J. Randers (1992). *Beyond the Limits*. Chelsea Green Publishing Company, Vermont.

Meadows, D. H., J. Randers, D. L. Meadows (2006). *Limits to Growth: The 30-Year Update*. Earthscan, London.

Meadows, D. L., J. Sterman, A. King (2020). Fishbanks: A renewable resource management simulation. https://mitsloan.mit.edu/LearningEdge/simulations/fishbanks/Pages/fish-banks.aspx (accessed May 2020).

Mella, P. (2007). *Systems Thinking Intelligence in Action*. Springer, Italy.

Moffat, I. (1991). *Causal and Simulation Modelling Using System Dynamics, Concepts and Techniques in Modern Geography*. Environmental Publications, Norwich.

Morecroft, J. D. W. (2015). *Strategic Modelling and Business Dynamics*. John Wiley & Sons Ltd., Sussex, U.K.

Muetzelfeldt, R., J. Massheder (2003). The simile visual modelling environment. *European Journal of Agronomy*, 18, 345–358.

Murray-Smith, D. J. (2015). *Testing and Validation of Computer Simulation Models: Principles, Methods and Applications*. Springer International Publishing, Switzerland.

Nirmalakhandan, N. (2002). *Modeling Tools for Environmental Engineers and Scientists*. CRC Press, Boca Raton, FL.

Oberkampf, W. L., C. J. Roy (2010). *Verification and Validation in Scientific Computing*. Cambridge University Press, New York.

Osais, E. Y. (2018). *Computer Simulation: A Foundational Approach Using Python*. CRC Press, Boca Raton, FL.

Press, W. H., B. P. Flannery, S. A. Teukolsky, W. T. Vetterling (1992). *Numerical Recipes in FORTRAN: The Art of Scientific Computing*. Cambridge University Press, Cambridge.

Richardson, G. P. (1986). Problems with causal loop diagrams. *System Dynamics Review*, 2(2), 158–170.

Richardson, G. P. (1996). *Modelling for Management: Simulation in Support of Systems Thinking*. Dartmouth Publishing Company, Aldershot.

Richardson, G. P. (1999). *Feedback Thought in Social Science and Systems Theory*. Pegasus Communications, Waltham, MA.

Richardson, G. P. (2011). Reflections on the foundations of system dynamics. *System Dynamics Review*, 27(3), 219–243.

Richardson, G. P., D. F. Andersen (2010). Systems thinking, mapping, and modeling for group decision and negotiation. *Handbook for Group Decision and Negotiation*, C. Eden, D. N. Kilgour, 313–324, Springer, Dordrecht.

Richardson, G. P., A. L. Pugh III (1981). *Introduction to System Dynamics Modeling* (1981). Productivity Press, Cambridge.

Richmond, B. (1993). Systems thinking: Critical thinking skills for the 1990s and beyond. *System Dynamics Review*, 9(2), 113–133.

Road Maps (2020). http://web.mit.edu/sysdyn/road-maps/D-4498.pdf (accessed May 2020).

Roberts, N., D. F. Andersen, R. M. Deal, W. A. Shaffer (1994). *Introduction to Computer Simulation. A System Dynamics Modeling Approach*. Productivity Press, New York.

Ross, S. (2013). *Simulation*. Elsevier, Amsterdam, The Netherlands.

Rossetti, M. D. (2016). *Simulation Modeling and Arena*. Wiley, New Jersey.

Ruth, M., B. Hannon (2009). *Dynamic Modeling of Diseases and Pests*. Springer, New York.

Ruth, M., B. Hannon (2012). *Modeling Dynamic Economic Systems*. Springer, New York.

Rutherford, A. (2020). *The Systems Thinker: Dynamic Systems*. Amazon, Inc.

Schwaninger, M., S. Grösser (2009). System dynamics modelling: Validation for quality assurance. *Encyclopedia of Complexity and Systems Science*, R. A. Meyers, Springer, New York. https://doi.org/10.1007/978-3-642-27737-5_540-4.

SD (2020). https://www.systemdynamics.org/what-is-sd (accessed May 20, 2020).

Senge, P. M. (1990). *The Fifth Discipline: The Art and Practice of the Learning Organization*. Doubleday Currency, New York.

Senge, P. M., C. Roberts, R. B. Ross, B. J. Smith, A. Kleiner (1994). *The Fifth Discipline Field Book: Strategies and Tools for Building a Learning Organization*. Doubleday Currency, New York.

Sherwood, D. (2002). *Seeing the Forest for the Trees: A Manager's Guide to Applying Systems Thinking*. Nicholas Brealey Publishing, London.

Simonovic, S. P. (2002). World water dynamics: Global modeling of water resources. *Journal of Environmental Management*, 66(3), 249–267.

Simonovic, S. P. (2008). *Managing Water Resources: Methods and Tools for a Systems Approach*. Earthscan, Routledge, New York.

Simonovic, S. P. (2010). *Systems Approach to Management of Disasters: Methods and Applications*. John Wiley & Sons, New York.

Sober, E. (2015). *Ockham's Razors: A User's Manual*. Cambridge University Press.

Soetaert, K., T. Petzoldt, R. W. Setzer (2010). *Solving Differential Equations in R*, 33, 9. https://EconPapers.repec.org/RePEc:jss:jstsof:v:033:i09 (accessed May 2020).

Sokolowski, J. A., C. M. Banks (2009). *Principles of Modeling and Simulation: A Multidisciplinary Approach*. Wiley, New Jersey.

Stave, K. A. (2003). A system dynamics model to facilitate public understanding of water management options in Las Vegas, Nevada. *Journal of Environmental Management*, 67, 303–313.

Sterman, J. D. (1984). Appropriate summary statistics for evaluating the historical fit of system dynamics models. *Dynamica*, 10(2), 51–66.

Sterman, J. D. (2000). *Business Dynamics. Systems Thinking and Modeling for a Complex World*. Irwin McGraw-Hil, New York.

Sterman, J. D. (2001). System dynamics modeling: Tools for learning in a complex world. *California Management Review*, 43(4), 8–25.

Sterman, J. D. (2002). All models are wrong: Reflections on becoming a systems scientist. *Systems Dynamics Review*, 18, 501–531.

Sweeney, L. B., D. H. Meadows (2010). *The Systems Thinking Playbook: Exercises to Stretch and Build Learning and Systems Thinking Capabilities*. Chelsea Green Publishing, Hartford, Vermont.

Teegavarapu, R. S. V. (2018a). Technology enhanced learning for civil engineering education: Use of dynamic and virtual-reality based simulation, online data analysis, and optimization tools. A. Rahman, V. Ilic *Blended Learning in Engineering Education: Recent Developments in Curriculum, Assessment, and Practice*, 305–320, CRC Press, Boca Raton, FL.

Teegavarapu, R. S. V. (2018b). *Trends and Changes in Hydroclimatic Variables: Links to Climate Variability and Change.* Elsevier, USA.

Teegavarapu, R. S. V., S. P. Simonovic (1999). Modeling uncertainty in reservoir loss functions using fuzzy sets. *Water Resources Research*, 35(9), 2815–2823.

Teegavarapu, R. S. V., S. P. Simonovic (2014). Dynamics of hydropower system operations. *Water Resources Management*, 28, 1937–1958. doi:10.1007/s11269-014-0586-2.

Thomas, H. A. (1981). *Improved Methods for National Water Assessment.* Report, Contract: WR15249270. US Water Resource Council, Washington, DC.

Vamvakeridou-Lyroudia, L. S., D. Savic (2008). System dynamics modelling: A tool for participatory simulation of complex water systems within aquas-tress. International Congress on Environmental Modeling and Software, 33, 2025–2028.

Vennix, J. A. M. (1999). *Group Model Building: Facilitating Team Learning Using System Dynamics.* John Wiley & Sons Ltd., Chichester, U.K.

Videira, N., P. Antunes, R. Santos (2009). Scoping river basin management issues with participatory modelling: The Baixo Guadiana experience. *Ecological Economics*, 68, 965–978.

Wolstenholme, E. F. (1990). *System Enquiry: A System Dynamics Approach.* John Wiley & Sons Ltd., New York.

Wolstenholme, E. F., R. G. Coyle (1983). The development of system dynamics as a methodology for system description and qualitative analysis. *Journal of the Operational Research Society*, 34(7), 569–581.

Wolstenholme, E. F., D. Mckelvie (2019). *The Dynamics of Care: Understanding People Flows in Health and Social Care.* Springer, Switzerland.

Chapter 2

Building Water and Environmental System Dynamics Simulation Models

Ramesh S. V. Teegavarapu

2.1 INTRODUCTION

System dynamics (SD) simulation concepts are introduced as descriptive modeling approaches for hydrologic modeling and water resources management problems in this chapter. The main generic building blocks (viz., stocks, flows, converters, and connectors) of the SD approach that can be used for the development of models for water and environmental systems are also discussed. Initially, a brief review of applications of SD models for water and environmental systems is provided in the next few sections.

2.1.1 Why SD Models for Water Resources Management

The field of water resources planning and management is replete with problems where traditional methods require a complex mathematical representation of the physical systems at the same time being excessively abstract. The inherent nonlinearities in hydrological processes and socio-economic aspects associated with water resources systems sometimes make them less amenable to traditional modeling approaches. Switching from reductionist approaches to holistic approaches may help to understand the dynamic behavior of these systems. SD simulation approach (Forrester, 1961) discussed in Chapter 1 can be effectively used to model many elements that govern the physical description and behaviors of water resource, environmental, and ecological systems.

An ideal example of a complex water resource system is a single or multiple-reservoir system. Management of any reservoir system is generally achieved by developing mathematical programming formulations for optimizing the operations or by using simulation models to understand or capture the dynamics of reservoir operation. Simulation models (Teegavarapu and Simonovic, 2000a) are often used in real time to refine the rules developed by optimization models (Wurbs, 1993; Teegavarapu, 2013; Teegavarapu and Simonovic, 2000b). Traditional simulation models developed using high-level computer programming languages or spreadsheet-based application software do not provide insights into the dynamics of the system behavior with time. Their structure is rigid, and the interaction between different components within the models is not transparent to the users. It is generally known that engineering systems are predictable and the dynamics of managed or engineered systems are well behaved and understood. Counter-intuitive system behavior that is common to natural, social, and economic systems is not usually observed in engineering systems. However, every water resource system is now evaluated from multiple social, ecological, and economic perspectives and therefore counter-intuitive system behavior is a possibility. Reservoir system operations are no exception to this as water as a resource is attached with a monetary value. The human dimension of the management of reservoirs governed by judgment, reliance on past experiences, and rules of thumb can at times lead to unpredictable or counter-intuitive behavior. This is true in the case of short-term or real-time operation of reservoir systems where the reservoir operators make changes to rules provided by optimization models in a short period. Random extreme events that influence the system also introduce a certain element of uncertainty that cannot be contemplated.

System dynamics concepts have been used in the past with the help of object-oriented simulation (OOS) environments to model different water resource systems. Several models can be envisioned in the field of environmental policy simulation studies where the different alternatives can be evaluated for efficient management strategies for planned irrigation districts. The advantage of using SD is that it allows the decision-maker to analyze different scenarios for a given management problem and adopt the best possible strategy that works over a period. It should be noted that SD is mainly a useful approach for understanding and simulating the behavior of systems and cannot be used for optimizing the operations of the systems

2.2 APPLICATIONS OF SD APPROACHES FOR MODELING WATER RESOURCES SYSTEMS

Processes influencing water resources systems can be understood by developing models using system dynamics (SD) principles. These models can be developed modeling environments referred to as object-oriented simulation (OOS) environments (Wurbs, 1993). Simulation environments that have drag-and-drop objects with specific properties can be used to develop these models easily. The use of OOS-based models utilizing SD principles for water resource planning and policy analysis studies (Lund and Ferreira, 1996; Simonovic et al., 1997; Fletcher, 1998; Simonovic and Fahmy, 1999) has shown benefits. Keyes and Palmer (1993) highlight the benefits of the SD approach to problems in water resources by demonstrating its utility in drought planning policy scenario generation. This approach would allow stakeholders and decision-makers to actively participate in the planning process and provide the facility to construct "what-if" scenarios. It also separates policy questions from data and makes the results and the model structure functionally transparent and acceptable to a wide group of parties involved in planning. Simonovic and Fahmy (1999) provide a general approach for policy analysis that uses the principles of system dynamics. Several studies (Ford, 1999; Deaton and Winebrake, 2000; Moffatt, 1991; Nirmalakhandan, 2002; Coyle, 1996) indicate that the principles of system dynamics are well suited for modeling and application to water resources and environmental problems. System behavior in space, as well as time, can also be simulated using a system dynamics framework (Huggett, 1993). Huang and Chang (2003) in a survey of tools for environmental systems analysis indicate the potential of using system dynamics for improved understanding of the environmental systems. The field of water resources has benefited from the system dynamics approach (Simonovic, 2009; Li and Simonovic, 2002). Simonovic et al. (1997) apply system dynamics simulation to the operation of the High Aswan Dam in Egypt. The advantages of using a simulation environment over developing a traditional simulation model are apparent from their study. Lund and Ferreira (1996) use the SD approach to develop a rule-based reservoir operation model. They compare the performance of this model with that of the Hydrologic Engineering Center-Prescriptive Reservoir Model (HEC-PRM).

A comprehensive treatise on the system dynamics approach to water resources management was provided by Simonovic (2008). Teegavarapu and Simonovic (2000a, 2014) discuss the use of SD principles and models for the evaluation of multi-reservoir operations. Ahmad and Simonovic (2000a; 2000b) used the SD approach to a reservoir operation problem for flood control. The impacts on flood management capacity of the reservoir are investigated by simulating a gated spillway in addition to an existing unregulated spillway. They indicate that the use of SD is more attractive than other techniques because of the ease of modification of system structure and the ability to perform sensitivity analyses. Saysel et al. (2002) evaluated water resources development options for regional agriculture using system dynamics simulation. Their emphasis was on the impact assessment of social and natural environments. Components affecting water demand and supply in the Yellow River basin in China were analyzed by Xu et al. (2002) using the SD approach. The value of water conservation was explained using an SD approach in a water management study in Las Vegas, USA, by Stave (2003). Several studies related to seawater intrusion (Fernandez and Selma, 2004), global water assessment (Simonovic, 2002), regional water resource assessment (Simonovic and Rajasekaram, 2004), flood evacuation strategies (Simonovic and Ahmad, 2005), and water conflict resolution (Nandalal and Simonovic, 2003) and hydrologic systems (Sehlke and Jacobson, 2005) have highlighted the advantages of using SD approach over traditional simulation methods. Simultaneous considerations of dynamic interactions between quantitative characteristics of water resources and water use constrained by socioeconomic, population, and physiographic heterogeneity were documented in multiple SD models (e.g., Simonovic and Rajasekaram, 2004; Simonovic, 2002) from the available literature. Applications of SD models for different areas in the water sector are discussed in several studies: water shortage assessments (Yang et al., 2008; Sušnik et al., 2012; Li et al., 2017) and sustainable water resources management (Winz et al., 2009; Simonovic, 2000; Sun et al., 2017; Madani and Mariño, 2009; Qin et al., 2011; Correia et al., 2019; Dawadi and Ahmad, 2013).

The planning and management of surface and subsurface hydrologic systems also provide enormous opportunities for the application of SD concepts. Simulation models can be developed for subsurface water assessment to evaluate several potential problems which include: (i) fate and transport of contaminants in a region; (ii) ground-level subsidence due to water extraction; (iii) pollutant migration; (iv) remediation alternatives; (v) risk assessment of any corrective actions taken; (vi) conjunctive use; and (vii) evaluation of recharge, pumping, and treatment strategies; some of these problems are addressed in studies by Barati et al. (2019). SD models can also be used to address several issues in the area of irrigation scheduling and planning. Several issues can be modeled which include: (i) irrigation scheduling and application options at the field level; (ii) simulation of soil moisture changes over time; and (iii) integrated reservoir–field-level models for the management of agricultural water allocations. SD models can also be used to investigate management alternatives associated with the conjunctive use of surface and subsurface water. Several policy options for groundwater management can be explored, and these options may not always be mutually exclusive. These options can be: (1) do nothing at this time; (2) permit groundwater management in areas where water extraction or pumping has adversely influenced streamflow; (3) augment water supplies wherever possible; and (4) adopt a different approach to conjunctive use. The final policy option would involve a change in the management of water use in urban areas, modification of soil conservation practices, and new measures for environmental enhancement.

Mashaly and Fernald (2020) have discussed numerous applications of SD for hydrologic modeling and water resources planning and management. They list several studies that dealt with water resources problems that are solved using the SD approach. These include groundwater modeling, surface water management, and modeling; holistic water resources management, water scarcity, and shortage; and water desalination and treatment, flood control, irrigation canal modeling, and water quality modeling. According to Mashaly and Fernanld (2020), studies have integrated the SD approach with other techniques and emerging methodologies and they are: (i) geographic information system (GIS); (ii) impact analysis (IA); (iii) agent-based modeling and simulation (ABMS); (iv) artificial intelligence (AI) and agent-based modeling (ABM); (v) exploratory modeling and analysis (EMA); (vi) analytic hierarchy process (AHP); (vii) fuzzy logic; (viii) Bayesian network (BN); (ix) artificial neural networks (ANNs); and (x) game theory (GT).

2.3 MODELING ENVIRONMENTAL SYSTEMS USING SD

Inductive and deductive modeling approaches have been commonly used to model and characterize environmental systems. Inductive approaches (data-driven or empirical) have also been used in many situations when the availability of water quality data is scarce or limited to characterize the system and develop conceptually acceptable deductive approaches. Inductive approaches (e.g., Reckhow and Chapra, 1983; Chapra 1994), process-based models (e.g., QUAL2E and HSPF), and comprehensive modeling environments (e.g., Better Assessment Science Integrating point and Non-point Sources (BASINS) environment) integrating a variety of process-based models were applied for modeling and assessment of pollutants in both surface and subsurface environments. These models are data-intensive and additional effort is required on the part of the modeler for calibration and validation of these models. Although these models attempt to overcome the limitations associated with other modeling approaches, their data requirements can be overwhelming. In many situations, the use of highly parameterized models may result in inconclusive results if enough data are not available. Craig et al. (2000) point to difficulties associated with the calibration of these models and the estimation of parameters from limited data.

In many situations, the parameters for these models cannot be uniquely obtained from the available field data and thus must be estimated from technical guidance documents (Bowie et al., 1985) or available literature. Often it is difficult to calibrate heavily parameterized process-based models with very little data, and the scientific credibility of the model and validation of model results is left to the judgment of the modeler. Also, little confidence can be attached to the results of the models (NRC, 2001) in such situations. On the other hand, simple inductive models can provide valuable insights into the processes without being highly parameterized (Craig et al., 2000). Considering these issues, researchers have advocated the development and use of conceptually simple models (e.g., Jian and Yu, 1998). The most useful predictive models are often extremely simple (Hodges, 1987; Levin 1985), and principles of SD (Nirmalakhandan, 2002; Deaton and Winebrake, 2000) are suited for the development of simulation models for environmental systems. SD approaches were also used for watershed modeling and management for nutrient and pathogen impairment of streams by Teegavarapu et al. (2005) and Elshorbagy et al. (2005, 2006). Teegavarapu et al. (2005) simulated the dynamics of a nutrient-impaired stream and evaluated source load reductions using a total maximum daily load (TMDL) approach. Applications of SD models for various water quality assessments are listed by Mashaly and Fernald (2020). A water quality simulation model was developed, calibrated, and validated using the SD approach. Details of this model are provided in Chapter 3.

2.4 SD BUILDING BLOCKS FOR WATER RESOURCE AND ENVIRONMENTAL SYSTEMS

Water resource systems are influenced by the spatial and temporal availability of water. Management of water resource systems will require a complete understanding of the hydrological cycle and all the interlinked components that are driven by different processes within that cycle. A simple mass balance equation that preserves the material (or mass) continuity between two spatial elements or in between two temporal windows is the fundamental equation to characterize water availability and movement within space and time. For example, the mass balance for a lake or a watershed can be carried out using the fundamental Equation 2.1 which indicates that change in the storage (ΔS) is equal to the difference in the two flow quantities (Inflows (I) – Outflows (O)).

$$\Delta S = I - O \tag{2.1}$$

A mass balance for a watershed with variables influencing the change in storage in any given time interval is given by Equation 2.2.

$$\Delta S = P - O + I - E - T - G \tag{2.2}$$

Where P: precipitation, O: surface runoff (or outflow), E: evaporation, T: transpiration (may be zero if there is no vegetation), and G: seepage of water into the groundwater. Different components of Equation 2.2 are expressed in not the same units of measurement. Unit conversions are required to make Equation 2.2 dimensionally homogeneous or consistent. The building blocks of SD discussed in Chapter 1 can be used to characterize any water resource or environmental system. Table 2.1 provides a list of components of the hydrologic and environmental systems and SD building blocks that can be used to model them. Dimensional homogeneity or consistency is critical to the model building using the SD approach. Therefore, all the flows and stocks identified when integrated within an SD model should be checked for consistency in units to achieve dimensional homogeneity.

The following sections will describe multiple steps in the modeling of water and environmental systems using SD principles. Causal loop diagrams (CLDs) are developed initially for each system identifying the main variables and feedbacks. A list of components or elements of each system that can be modeled as stocks and flows, converters, and other objects is provided. Then using these objects, stock and flow diagrams (i.e.,

Table 2.1 SD Building Blocks That Can Be Used to Model Major Components of a Hydrologic System

SD Building Block	Component of a Hydrologic System
Stocks	Water stored on land surface (reservoirs, lakes, ponds)
	Interception (water intercepted by any object)
	Depression storage (water stored in depressions, puddles)
	Snow accumulation
	Root zone storage (water stored at the root zone, subsurface storage)
	Groundwater storage (subsurface water)
	Channel storage (water stored temporarily in the channel)
Flows	Precipitation
	Evaporation
	Transpiration
	Surface runoff (overland flow)
	Channel flow
	Infiltration (into the ground)
	Channel seepage
	Percolation (infiltration into the deep groundwater zone)
	Groundwater flow
	Evapotranspiration
	Baseflow
	Diversion flows from reservoirs or channels
	Controlled releases from reservoirs
	Releases from detention ponds

structure diagrams) developed using the STELLA modeling environment are presented.

2.4.1 Model Development Steps

Development of SD models for water (i.e., hydro) and environmental systems can be carried out using a series of steps:

- Identify the main processes and variables that govern the behavior of the hydro-environmental system under consideration.
- Describe the system qualitatively first by understanding causal relationships between different hydroclimatic and environmental variables.
- Collect all relevant data (i.e., hydrological, meteorological, climatological, and environmental) related to the system.
- Define the boundaries of the system by identifying the spatial extent.
- Define the temporal horizon over which the system is expected to be simulated.

- Identify the feedbacks and any human influences (i.e., hydroclimatic and environmental variables influenced by human activities) and interventions (i.e., operation or management of hydrosystems) that affect the behavior of the system.
- Develop a causal loop diagram (CLD) and identify the dominant loop(s) in the system.
- Identify stocks (e.g., reservoirs and lakes) and flows (i.e., natural inflows and releases from reservoirs).
- Select an SD simulation model development environment with necessary features and functionalities.
- Create a stock and flow diagram with all the connections established between different components of the system in an SD simulation model development environment.
- Determine the temporal aspects of the simulation including the model time horizon (i.e., day, week, and year) and time step (i.e., sub-daily interval) and select a numerical integration scheme (i.e., Euler or one of the two variants of Runge-Kutta (RK2 and RK4)).
- Divide or partition the available system-related data into two sets: data for calibration and validation.
- Calibrate the SD model using available historical data by tweaking or adjusting parameter values and confirm the system behavior with real-life conditions to obtain appropriate values of parameters. Error measures and performance measures can aid in the calibration exercise.
- Verify and validate the SD model using the data set aside and evaluate the performance of the model in accurately reproducing the system behavior in all possible conditions.
- Re-calibrate and validate with additional datasets or restructure the SD model considering additional variables and relationships if the validation results are not acceptable.
- Simulate the system behavior for policy analysis after successful calibration and validation.

The following sections provide details of SD model development for different water and environmental systems. For each system, causal loop diagrams (CLDs), building blocks (stocks, flows, converters, connectors) of the SD approach, and stock and flow diagrams (Keating, 2020) developed using the STELLA modeling environment are presented. CLDs developed and presented in this chapter adhere to most of the rules discussed in Chapter 1. Minor deviations from those standard rules for CLD development may be noted. There might be some restrictions imposed by the STELLA modeling environment when stock and flow diagrams are developed and ultimately used for simulation of the systems. In almost all models discussed in this chapter, the system boundaries are defined by sources and sinks identified as objects depicted as clouds in STELLA. Also, it should be noted that flows may exist independently (i.e., without being connected to stocks) in many stock and flow diagrams that are related to water and environmental systems. In these flows that are not connected to stocks, time series of variables (e.g., streamflows) can be represented.

2.4.2 Modeling Environment

The STELLA modeling environment used to develop stock and flow diagrams discussed in this chapter provides several features for the development of SD models. A brief overview of the features available in this environment is provided in this section. Drag-and-drop objects (i.e., stocks, flows, converters, and connectors) are available to pull them on to a canvas to build models. Graph and table objects can also be pulled onto the canvas to display changes in variables using time series plots and list the numerical values of the variables, respectively. Time series values of multiple objects (i.e., stocks, flows, and converters) simultaneously can be visually evaluated. Objects can be deleted in any part of the model and exact replicas of the objects (referred to as ghosts) can be made. The ghosts can be strategically placed in the model when required to avoid clutter and long overlapping connectors from one part of the model to another. The model can be divided into parts and each part can be designed as a sector. Each sector can be run independently. Alerts are provided to the environment whenever impossible variable values are realized during the simulation. Three integration methods (viz., Euler, Runge-Kutta 2 (RK2), and Runge-Kutta 4 (RK4)), a mechanism to change the simulation time interval, start and end times, and simulation speed are available to the modelers. These selections are available under the run specifications menu of the modeling environment. Several built-in functions such as mathematical, trigonometric, logical, statistical, financial, discrete, cycle-time are available. The sensitivity analysis feature provides the user to perturb a variable value within a specific range using uniform and Gaussian probability distributions. This feature can help to analyze the system with Monte Carlo (MC) simulation. STELLA provides users with the ability to introduce variations in variables using different variable perturbation options which include incremental, distributions, ad hoc, and user-specified datasets. Users can conduct sensitivity analysis using the *Sensi* specification feature available in the simulation environment. The modeling environment also includes a feature that helps to check the unit consistency of variables. A flight simulation mode allows the users to stop the simulation at any time and make changes to the variables.

2.5 SINGLE-RESERVOIR SYSTEM

The causal loop diagram (CLD) for a reservoir operation problem is shown in Figure 2.1. The following observations can be made from the

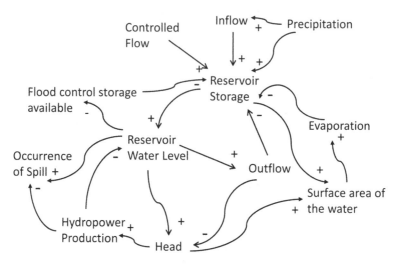

Figure 2.1 Causal loop diagram for reservoir operation problem.

CLD. Reservoir storage (i.e., the amount of water stored in the reservoir) increases as water contributing sources (flows) to the reservoir increase. The flows include natural flow or inflow (generally referred to as streamflow), controlled flow (flow controlled by hydrologic and hydraulic structures), and precipitation. The increase in the volume of water in the reservoir also increases the water level in the reservoir. The polarity sign "+" indicates this cause and effect relationship.

As the reservoir storage increases, the water level increases which in turn increases the surface area of the water in the reservoir. An increase in the surface area of the water increases evaporation. An increase in reservoir level triggers or increases the outflow as the level and discharge (flow) from the reservoir are interconnected by the height of water (head) in the reservoir. An increase in the occurrence of the spill and the reduction in the flood control storage (storage set aside for handling floods) is possible due to an increase in the level of the water in the reservoir. Finally, if hydropower is produced using the reservoir, then an increase in the head increases the amount of power that can be produced. Since the reservoir operation represents a managed system for the storage and distribution of water for multiple purposes, the system behavior is governed by negative feedback with human intervention. The exogenous shocks (i.e., rare record-breaking floods or droughts) to the reservoir system may lead to positive feedback behavior dominating the system for a short period.

The generic building blocks of SD can be used to build a model using stocks, flows, converters, and connectors (i.e., links). The components of the reservoir system and the corresponding building blocks are provided in Table 2.2. Functional relationships between surface area and (1) storage, (2)

Table 2.2 Building Blocks for the SD Model for the Reservoir Operation

SD Building Blocks	Component(s) of the System
Stock(s)	Reservoir storage
	Flood control storage
Flow(s)	Inflow
	Outflow
	Precipitation
	Evaporation
	Controlled flow
	Spill
Convertor(s)	Head
	Hydropower production
	Surface area of the water
Source(s)	Origin of inflow, controlled flow, precipitation
Sink(s)	Destination of spill, evaporation, outflow

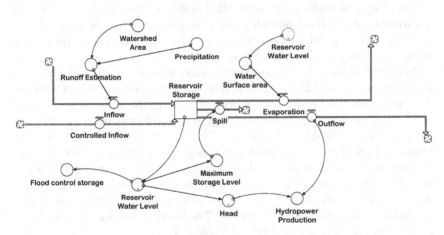

Figure 2.2 Stock and flow diagram for reservoir operation model.

evaporation rate, and (3) water level are needed to develop this model. Time series of datasets related to flow values (both controlled and inflows), precipitation is required. Information about flow released (outflow), the head, and details of types of turbines used and efficiency as a function of discharge are required to estimate the amount of hydropower produced (Figure 2.2).

2.6 WATER BALANCE

The causal loop diagram presented in Figure 2.3 shows different components of the mass balance of a watershed system considering surface (above the ground) and subsurface (below the ground) storage components. This is

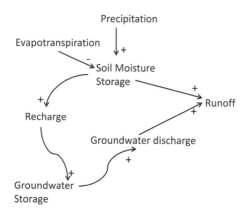

Figure 2.3 Causal loop diagram for the water balance of a region.

a natural system without any controlled flows or managed flows. The soil moisture zone in a fixed bounded region of the subsurface is modeled as a quantity that accumulates or decreases based on different flows (referred to as inputs or outputs) into the system.

Precipitation increases the amount of water available in the soil zone (i.e., soil moisture storage) and evapotranspiration. Evaporation (loss of water to the atmosphere) and transpiration (loss of water to the atmosphere through vegetation) are possible as roots existing in this soil moisture zone decrease the amount of available soil moisture. The polarity signs for evapotranspiration and precipitation are provided as negative and positive, respectively. An increase in soil moisture leads to an increase in recharge (flow of water into deeper groundwater systems) and an increase in groundwater storage. An increase in groundwater storage and soil moisture combined with precipitation contributes to higher runoff. The building blocks of SD and corresponding components are listed in Table 2.3.

Table 2.3 Building Blocks for the SD Model for the Water Balance of a Region

SD Building Blocks	Component(s) of the System
Stock(s)	Soil moisture storage, groundwater
Flow(s)	Recharge
	Runoff
	Precipitation
	Evapotranspiration
	Groundwater discharge
Source(s)	Origin of precipitation
Sink(s)	Destination of evapotranspiration, runoff

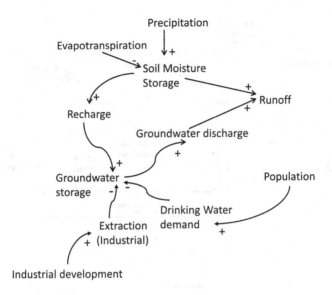

Figure 2.4 Causal loop diagram for the water balance of a region with additional components.

A causal loop diagram for the water balance model with additional components (i.e., population, drinking water demand, industrial extraction, and industrial development) can also be developed as shown in Figure 2.4. Feedbacks for these components indicate that increase in the population would lead to an increase in drinking water demands; this, in turn, may decrease groundwater storage if water needed is obtained from groundwater resources. Also, the increase in industrial development leads to an increase in the extraction of groundwater. The stock and flow diagrams for the basic water balance model and with additional components are shown in Figures 2.5 and 2.6, respectively.

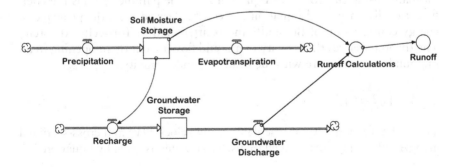

Figure 2.5 Stock and flow diagram for water balance model.

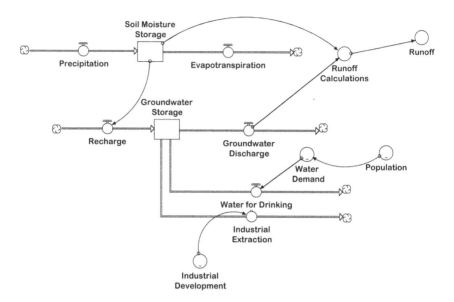

Figure 2.6 Stock and flow diagram for water balance model with additional components.

2.7 ROUTING OF FLOWS THROUGH DETENTION POND SYSTEM

Detention ponds are mainly used as one of the structural measures adopted for urban storm runoff (or stormwater) management (Behera and Teegavarapu, 2015; Akan and Houghtalen, 2003). These ponds (small reservoirs or impoundment structures) constructed in urban areas will temporarily store or impound the water and let it move out of the system through outlet(s). An outlet is a pipe that carries water from the pond to the nearby discharging point (lake, river, or canal). The magnitude of the release from the outlet is mainly controlled by the head (i.e., the height of water above the outlet or opening). The velocity (V) of water through the outlet can be estimated using the Torricelli Equation 2.3. The parameter C_d is the coefficient of discharge and Equation 2.3 can be derived from the principle of energy conservation of Bernoulli and is attributed to Torricelli's theorem. The theorem relates the velocity of a liquid flowing out of an opening in an impoundment structure with the head (H) under gravity.

$$V = \left(\sqrt{2gH} \right) \tag{2.3}$$

The outlet discharge (Q) is calculated using velocity (V) and cross-sectional area (A_p) of the pipe that is used to discharge water as given by Equation 2.4.

$$Q = C_d V A_p \tag{2.4}$$

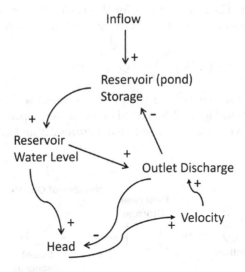

Figure 2.7 Causal loop diagram for routing of flows in the detention pond.

The CLD for the detention pond system is provided in Figure 2.7. The polarity signs are placed to indicate cause and effect relationships. The pond storage increases as the inflows (storm runoff) increase and contributes to an increase in the water level in the pond. The increase in the water level influences the head and increases the velocity of water through the pipe, hence the release (or outlet discharge). The increase in the discharge decreases the water level, and the system is self-controlled for design conditions that allow accommodating water in the pond. The blocks (stocks and flows) that can be used for the development of the model for this system and corresponding components are listed in Table 2.4. The outlet discharge is based on a storage–discharge relationship and can be included in the SD model

Table 2.4 Building Blocks for the SD Model for Routing Flows in the Detention Pond

SD Building Blocks	Component(s) of the System
Stocks	Reservoir(pond) storage
Flows	Inflow
	Outlet discharge
Convertor(s)	Reservoir water level
	Head
	Velocity
	Storage–discharge relationship
Source(s)	Origin of precipitation
Sink(s)	Destination of outlet discharge

using a converter. The routing Equation 2.5 is based on the mass balance in two intervals (t and $t + 1$) and is solved for each time interval.

$$(I_t + I_{t+1}) + \left(\frac{2S_t}{\Delta t} - Q_t\right) = \left(\frac{2S_{t+1}}{\Delta t} + Q_{t+1}\right) \quad \forall t \tag{2.5}$$

The stock and flow diagram for the detention pond with additional components is shown in Figure 2.8. The SD models developed in STELLA and Insight Maker environments are shown in Figures 2.9 and 2.11, respectively.

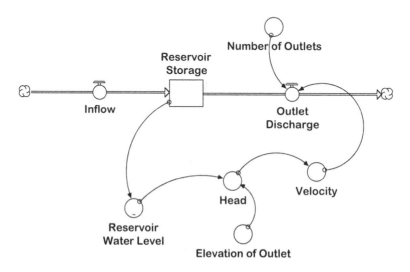

Figure 2.8 Stock and flow diagram for a detention pond.

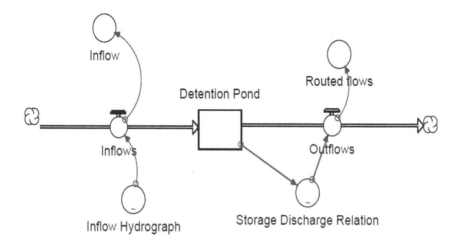

Figure 2.9 Detention pond routing model developed using STELLA.

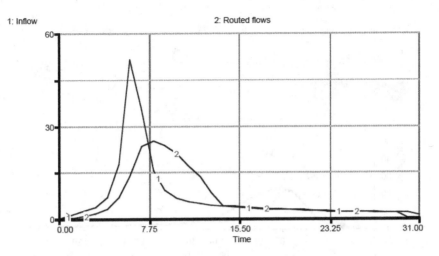

Figure 2.10 Variation of inflow and routed flows over time based on the STELLA interface.

Converters are used to obtain information about inflow and routed flows (outlet discharges). They are also used to define two relationships: (1) time and inflow (i.e., inflow hydrograph) provided as a graphical relationship and (2) storage–discharge (outlet discharge) relationship. The sources and sinks for this system are shown as system boundaries (Figure 2.9). Results from simulation model using STELLA are shown in Figure 2.10.

Details of the detention pond routing modeled using Insight Maker and results of routing the inflows are shown in Figures 2.11 and 2.12, respectively.

The inflows to the reservoir that are routed are based on post-development conditions. The hydrograph associated with pre-development conditions is also estimated and is required to check the success of the detention

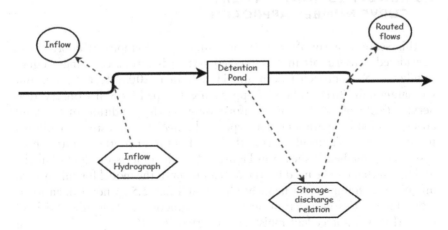

Figure 2.11 Detention pond routing model developed using Insight Maker.

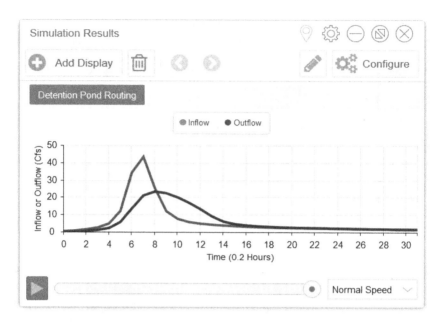

Figure 2.12 Variation of inflow and routed flows over time based on Insight Maker.

pond design. If the peak discharge associated with the routed hydrograph through the pond is higher than the pre-development hydrograph then two options are available and they are: (1) increase the volume of the pond and (2) decrease the number of the outlets or discharge from the outlet by reducing the diameter of the pipe that serves as an outlet.

2.8 RUNOFF ESTIMATION USING CURVE NUMBER APPROACH

In this section, a runoff estimation using a semi-empirical approach is considered. The runoff in a watershed system is estimated using a curve number (CN) approach (Akan and Houghtalen, 2003; Teegavarapu and Chinatalapudi, 2018). The CN approach developed by Soil Conservation Service (SCS) of the U.S. uses information on the condition of the land cover, precipitation amount, soil type, and antecedent moisture conditions to estimate the magnitude of runoff. The CLD for a runoff generation process in a watershed is shown in Figure 2.13. The objects or blocks (stocks and flows) that can be used for the development of the model for this system and corresponding components are listed in Table 2.5. A non-dimensional CN which varies between 40 and 98 is a function of land use and land cover (LULC), antecedent moisture conditions (AMC) or pre-existing soil moisture conditions, hydrological soil type or group, extent, and nature of

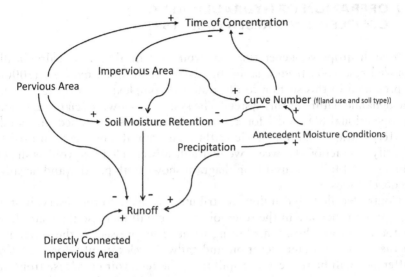

Figure 2.13 Causal loop diagram for runoff estimation.

Table 2.5 Building Blocks for the SD Model for Runoff Estimation

SD Building Blocks	Component(s) of the System
Stock(s)	Soil moisture retention
Flow(s)	Precipitation
	Runoff
Convertor(s)	Time of concentration
	Pervious area
	Impervious area
	Curve number
	Directly connected impervious area
Source(s)	Origin of precipitation
Sink(s)	Destination of runoff

the impervious area, and time of concentration. As the impervious area in the watershed or antecedent moisture increases, the CN number based on soil moisture retention capacity for watershed and runoff increases. The moisture retention ability of soil decreases and therefore runoff increases for a storm event. The decrease in time of concentration can be attributed to the increase in impervious area. Also, an increase in the pervious area in the watershed results in increase in the time of concentration and runoff. An increase in the direct connected impervious area (impervious area directly connected to the outlet of the watershed) increases runoff and reduces the time of concentration.

2.9 OPERATION OF HYDRAULICALLY COUPLED TWO-RESERVOIR SYSTEM

A two hydropower-generating reservoir system that has hydraulically coupled reservoirs from a study by Teegavarapu and Simonovic (2000b) is presented in this section to illustrate the complexities of operation and the dynamic nature of this system. The system is shown in Figure 2.14 and an operational SD model for a similar system was developed in a study by Teegavarapu and Simonovic (2014). The causal loop diagram used to identify interactions between two hydraulically coupled reservoirs is shown in Figure 2.14. The causal loop diagram shows both positive and negative feedback loops.

Controlled flow and inflow contribute positively to the reservoir storage, and an increase in the reservoir storage will increase the water level in the reservoir, thus contributing to head (expressed as the difference in the reservoir water elevation and tailwater elevation). However, this difference will be reduced if a spill from the reservoir or release from the reservoir occurs thus reducing the head. Also, in a hydraulically coupled reservoir system as shown in Figure 2.14, any addition of inflow, local inflow, release, or spill from the first reservoir will increase the downstream reservoir storage and increase the reservoir water level of the second reservoir. This increase in the reservoir level increases the tailwater elevation and thus reduces the head. The tailwater elevation is also reduced when release is made from reservoir 2. The CLD for a runoff process in a watershed is shown in Figure 2.15.

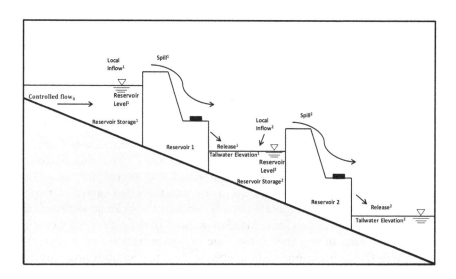

Figure 2.14 Schematic of a two-reservoir hydropower system with hydraulic coupling.

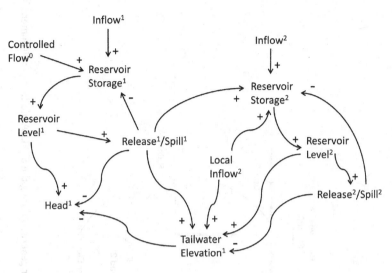

Figure 2.15 Causal loop diagram explaining positive and negative feedbacks in a two-reservoir system used for hydropower generation.

Table 2.6 Building Blocks for the SD Model for the Two-Reservoir System

SD Building Blocks	Component(s) of the System
Stock(s)	Reservoir storages
Flow(s)	Local inflow
	Controlled inflow
	Release
	Spill
Convertor(s)	Reservoir levels
	Head
	Tailwater elevation
Source(s)	Origin of: local inflow, controlled inflow
Sink(s)	Destination of: release, spill

The blocks (stocks and flows) that can be used for the development of the model for the two-reservoir system and corresponding system components are listed in Table 2.6. The stock and flow diagram for the reservoir system is shown in Figure 2.16.

2.10 POLLUTANT MANAGEMENT IN A STREAM

A pollution management problem is addressed using an SD model in this example. The CLD for a stream that is impaired due to pollutants from

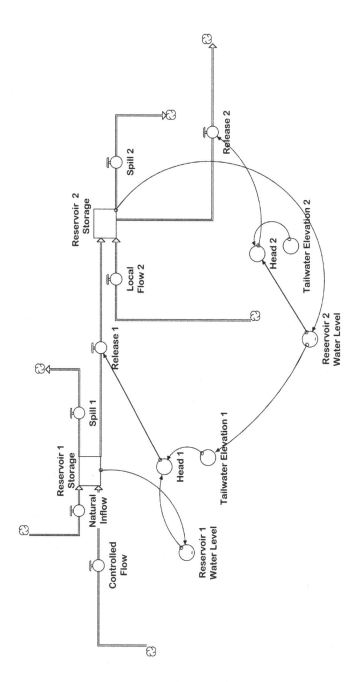

Figure 2.16 Stock and flow diagram based on a two-reservoir system used for hydropower generation using the STELLA modeling environment.

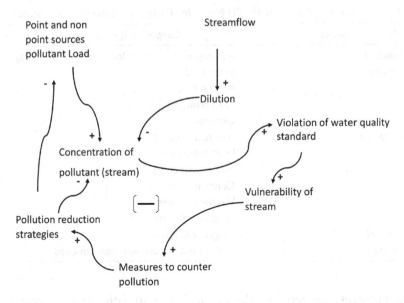

Figure 2.17 Causal loop diagram for pollutant management problem.

multiple point and non-point sources is shown in Figure 2.17. The pollutant loads from different sources when they are added to the stream increase the concentration of the pollutant in the stream. Pollutant concentration is calculated based on the amount of pollutant load and flow (discharge) in the stream. However, when more water joins, the concentration of the pollutant decreases due to dilution. An increase in the pollutant concentration increases the level of impairment of the stream. To manage pollution, different management strategies are implemented to reduce pollution in the stream. The management strategies are also used to reduce the contributions from pollutant sources. The blocks (stocks and flows) that can be used for the development of the model for this system and corresponding system components are listed in Table 2.7. Figure 2.18 shows the stock and flow diagram for pollutant management in a stream using the STELLA modeling environment.

2.11 POLLUTANT MANAGEMENT IN A WATERSHED

Pollutants entering streams in a watershed due to point and non-point sources can be modeled using the SD approach. An example of pathogen (i.e., fecal coliform) impairment of streams due to non-point sources in a watershed is discussed in this section. More details about this specific problem using the SD approach are reported by Teegavarapu et al. (2002). The CLD describing different components of the system is shown in Figure 2.19.

Table 2.7 Building Blocks for the SD Model for Pollutant Management in a Stream

SD Building Blocks	Component(s) of the System
Stock(s)	Accumulated pollutant load, water body storage
Flow(s)	Inflows
	Pollutant load
	Load removed
	Outflows
Convertor(s)	Pollutant sources
	Load reduction
	Impairment check
	Concentration
	Volume of water
	Water quality standard
Source(s)	Origin of inflows
Sink(s)	Destination of outflows, load removed

Point sources considered are those contributing to pathogen (i.e., fecal coliform or E-coli) loading in streams via discharge of the human and animal waste directly into streams without treatment. The decay of pathogens over time is modeled using a first-order decay process, and pathogen load and flow relationships are used to obtain the loads that are driven by rainfall related events. A loading rate is calculated based on the number of point sources, and a reduction parameter associated with a decrease of sources is included in the model to evaluate pollution reduction strategies to reduce pathogen impairment of streams. The blocks (stocks and flows) that can be used for the development of the pollutant management model and the corresponding components are listed in Table 2.8. The stock and flow diagram for this pathogen impairment management problem is shown in Figure 2.20.

2.12 DATA REQUIREMENTS FOR SD MODELS

The development of SD models for water and environmental systems described in this chapter also require different types of datasets. These include flow (or streamflow) data as time series, initial values of storage elements, values of parameters, and constants related to different functional relationships that quantify the values of different components of the system. Functional relationships that relate different variables, graphical relationships that link two variables, conditions required for rules to specify flows, delays, and material transfers are required for the creation of SD models. Calibration and validation of the models are two critical tasks that need to be carried out before the models can be used for a policy or scenario-based

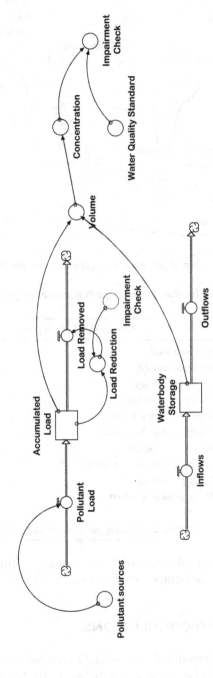

Figure 2.18 Stock and flow diagram for pollutant management using the STELLA modeling environment.

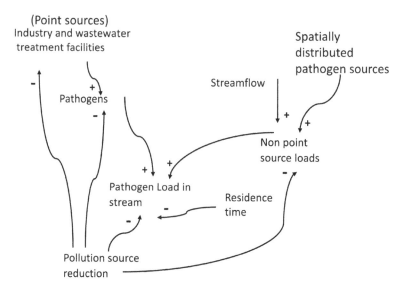

Figure 2.19 Causal loop diagram of pollutant management in a watershed.

Table 2.8 Building Blocks for the SD Model for Pollutant Management in a Watershed

SD Building Blocks	Component(s) of the System
Stock(s)	Pathogen load
Flow(s)	Pathogen loading
	Surviving pathogen load
	Dead pathogen load
Convertor(s)	Streamflow time series
	Pathogen load flow relationship
	Number of sources
	Source reduction parameter
	Decay parameter
	Water quality standard
Source(s)	—
Sink(s)	Destination of surviving pathogen load, dead pathogen load

analysis. Manual (e.g., trial and error) or automatic optimization procedures are required for the estimation of optimal parameters of the model.

2.13 SUMMARY AND CONCLUSIONS

The development of SD models for water and environmental systems is discussed in this chapter. The rationale for the use of the SD approach and a

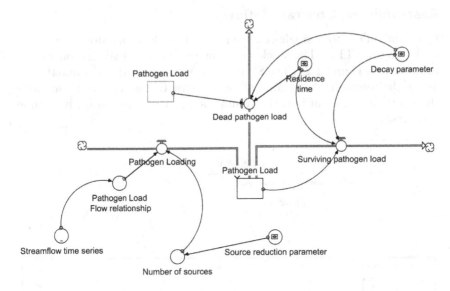

Figure 2.20 Stock and flow diagram of the pathogen model.

few applications of the same are also discussed in this chapter. The creation of causal loop diagrams (CLDs) for different water and environmental systems is discussed along with the identification of major building blocks that can be used to represent and model different components of these systems in proprietary and public simulation environments. Stock and flow diagrams using a simulation modeling environment are also provided to help understand the transition from CLDs to SD simulation models. Different systems described in this chapter will no way cover all the aspects of water and environmental systems and any connected economic and social systems. Readers are referred to several studies listed in this chapter for further exploration of applications of SD approaches to water and environmental systems modeling and management. The next chapter will describe two case studies that adopt SD models for modeling and management of water and environmental systems.

APPENDIX: RESERVOIR SYSTEM DYNAMICS SIMULATION MODEL

A single-reservoir operation model developed using STELLA is described in this section. An SD model with one inflow is extended using additional outflows and other variables as described in the next several sections. The units for different variables (i.e., stocks and flows and others) are avoided to keep the model generic.

Reservoir and Constant Inflow

Constant inflow to a single-reservoir system with initial storage is modeled using STELLA. The stock and flow diagram and simulation results are shown in Figure 2.21. The SD model is simulated using the initial storage of the reservoir of 100 and inflow value of 10. As there is no outlet to the reservoir or no outflow, the volume of water increases linearly as time progresses.

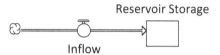

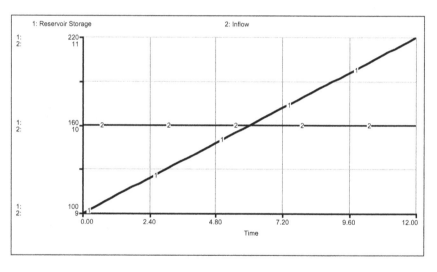

Figure 2.21 Stock and flow diagram and simulation of a reservoir operation with a constant value of inflow.

Reservoir and Varying Inflow

Varying inflows that are cyclical (shown in Figure 2.22) to a reservoir with initial storage is modeled using STELLA. The stock and flow diagram and simulation results are shown in Figure 2.23. The SD model is simulated using the initial storage of the reservoir of 100 units and inflow value of 10. As there is no outlet to the reservoir or no outflow, the volume of water increases nonlinearly as time progresses based on the variation of the inflows. The rate of increase in storage is higher in time intervals when larger inflows occur compared to that of those with lower inflows.

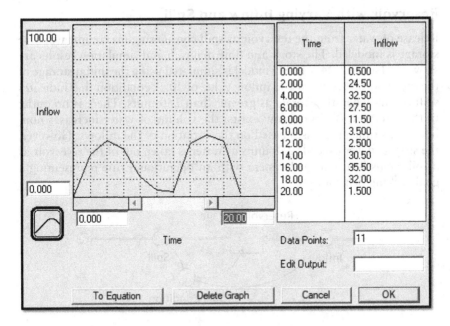

Time	Inflow
0.000	0.500
2.000	24.50
4.000	32.50
6.000	27.50
8.000	11.50
10.00	3.500
12.00	2.500
14.00	30.50
16.00	35.50
18.00	32.00
20.00	1.500

Data Points: 11

Edit Output:

To Equation   Delete Graph   Cancel   OK

Figure 2.22 Varying inflows used in the simulation.

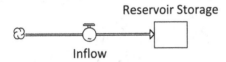

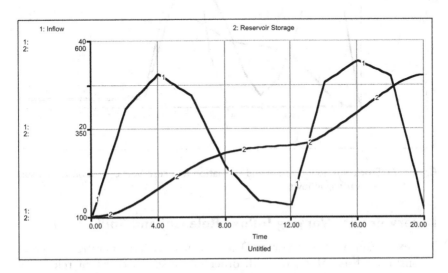

Figure 2.23 Stock and flow diagram and simulation of a reservoir operation with varying inflows.

Reservoir with Varying Inflow and Spill

Release of water from the reservoir as spill considering maximum available storage is modeled. The stock and flow diagram and simulation results are shown in Figure 2.24. The SD model is simulated using the initial storage of the reservoir of 100 units and inflow value of 10. A condition that indicates spill occurs when the storage is greater than 250 units. There is no outlet to the reservoir, or no outflow exists; the volume of water increases non-linearly as time progresses based on the variation of the inflows. However, the storage never exceeds 250 units as water is taken out of the reservoir as a spill. This is a managed system with an operational rule representing a goal-seeking behavior.

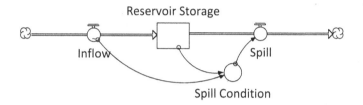

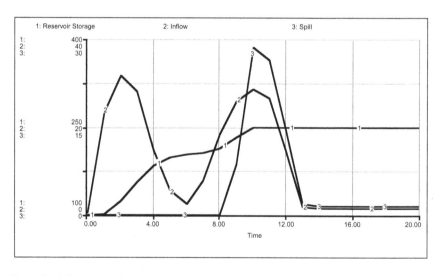

Figure 2.24 Stock and flow diagram and simulation of a reservoir operation with a spill and varying inflows.

Reservoir with Varying Inflow, Release, and Spill

The extension of the previous SD model of reservoir operation includes a constant release along with all other processes. A constant release of 10 units is included in this model suggesting another outflow combined with condition outflow associated with the spill. The stock and flow diagram and simulation results are shown in Figure 2.25. The SD model is simulated using the initial storage of the reservoir of 100 units and inflow value of 10.

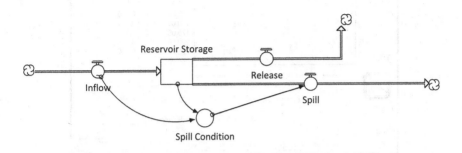

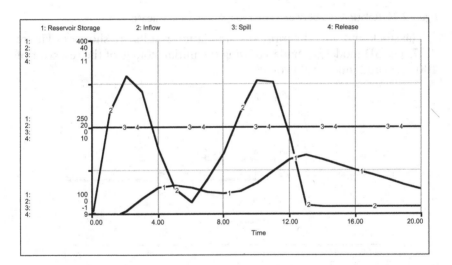

Figure 2.25 Stock and flow diagram and simulation of a reservoir operation with a spill, constant release, and varying inflows.

Reservoir with Varying Inflow, Conditioned Release, and Spill

In this SD model, along with varying inflow and spill, a release amount conditioned on the storage in the reservoir is included. A nonlinear relationship (shown in Figure 2.26) relating to reservoir storage and release (referred to as a release condition) is incorporated in the model. This relationship reflects an uncontrolled outlet to a reservoir and the discharge from this outlet depends on the water volume in the reservoir.

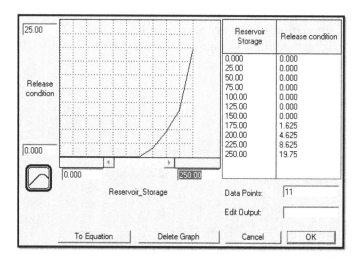

Figure 2.26 Relationship between reservoir storage and release.

The stock and flow diagram and simulation results are shown in Figure 2.27. The SD model is simulated using the initial storage of the reservoir of 100 units and inflow value of 10.

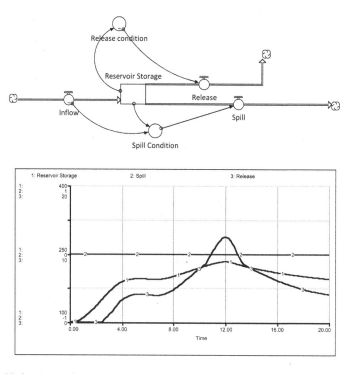

Figure 2.27 Stock and flow diagram and simulation of a reservoir operation with a spill, conditioned release, and varying inflows.

Reservoir with Varying Inflow, Conditioned Release, Evaporation Loss, and Spill

The SD model now incorporates evaporation from the water surface in the reservoir as a loss in any given time interval. Evaporation depends on the surface area and increases with the increase in the latter. As the volume of the water increases in the reservoir, the surface area occupied by the water also increases in general. A nonlinear relationship between reservoir volume and surface area is used in this model as shown in Figure 2.28. Similarly, a relationship between surface area and evaporation as shown in Figure 2.29 is incorporated in the model. The evaporation loss from the reservoir is thus estimated using two relationships (i.e., reservoir storage vs surface area and surface area vs evaporation) is modeled as an outflow.

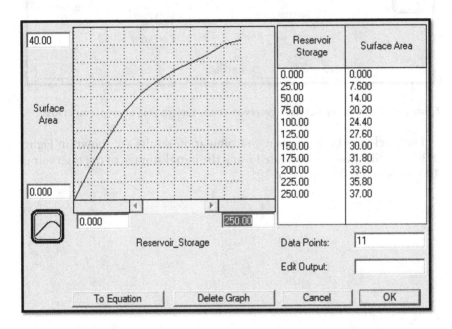

Figure 2.28 Relationship between reservoir volume and surface area of the water.

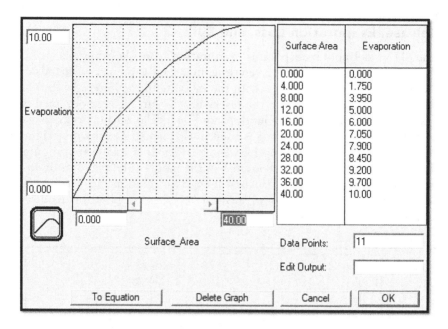

Figure 2.29 Relationship between reservoir water surface area of evaporation loss.

The stock and flow diagram and simulation results are shown in Figure 2.30. The SD model is simulated using the initial storage of the reservoir of 100 units and inflow value of 10.

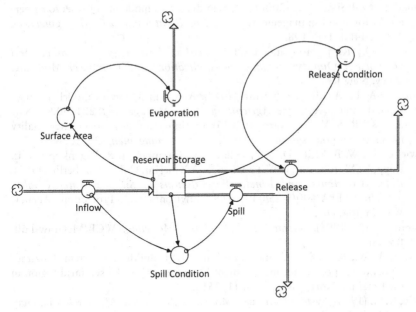

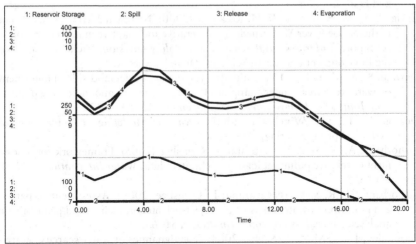

Figure 2.30 Stock and flow diagram and simulation of a reservoir operation with spill, release, evaporation, and varying inflows.

REFERENCES

Ahmad, S., S. P. Simonovic (2000a). Dynamic modeling of flood management policies. *Proc. of the 18th International Conference of the System Dynamics Society, Sustainability in the Third Millennium*, 1–10, Pal I. Davidsen, David N. Ford, Ali N. Mashayekhi, Elizabeth F. Andersen, Bergen, Norway.

Ahmad, S., S. P. Simonovic (2000b). System dynamics modeling of reservoir operations for flood management. *Journal of Computing in Civil Engineering, ASCE*, 14(3), 190–198.

Akan, A. O., R. J. Houghtalen (2003). *Urban Hydrology, Hydraulics, and Stormwater Quality: Engineering Applications and Computer Modeling.* Wiley, New Jersey.

Barati, A. A. H. Azadi, J. Scheffran (2019). A system dynamics model of smart groundwater governance. *Agricultural Water Management*, 221, 502–518.

Behera, P. K., R. S. V. Teegavarapu (2015). Optimization of a stormwater quality management pond system. *Water Resources Management*, 29, 1083–1095.

Bowie, G. L., W. B. Mills, D. B. Porcella, C. L. Campbell, J. R. Pagenkopf, G. L. Rupp, K. M. Johnson, P. W. H. Chan, S. A. Gherini, C. E. Chamberlin (1985). *Rates, Constants, and Kinetic Formulations in Surface Water Quality Modeling.* EPA/600/3-85/040, U.S. Environmental Protection Agency, Washington, DC.

Chapra, S. C. (1994). *Surface Water Quality Modeling.* WCB/McGraw-Hill, Boston.

Correia, A. W., K. P. E. Oliveira, O. Sahin (2019). Building a system dynamics model to support water management: A case study of the semiarid region in the Brazilian Northeast. *Water*, 11, 2513.

Coyle, R. G. (1996). *System Dynamics Modeling: A Practical Approach.* Chapman and Hall, London.

Craig, A. S., M. E. Borsuk, K. H. Reckhow (2000). Nitrogen TMDL development in the Neuse River Watershed: An imperative for adaptive management, project report. *Environmental Sciences and Policy Division*, Nicholas School of the Environment and Earth Sciences, Duke University.

Dawadi, S., S. Ahmad (2013). Evaluating the impact of demand-side management on water resources under changing climatic conditions and increasing population. *Journal of Environmental Management*, 114, 261–275.

Deaton, M. L., J. I. Winebrake (2000). *Dynamic Modeling of Environmental Systems.* Springer-Verlag, New York.

Elshorbagy, A., R. S. V. Teegavarapu, L. Ormsbee (2005). Framework for assessment of relative pollutant loads with limited data. *Water International*, 30(3), 350–355.

Elshorbagy, A., R. S. V. Teegavarapu, L. Ormsbee (2006). Assessment of pathogen pollution in watersheds using object-oriented modeling and probabilistic analysis. *Journal of Hydroinformatics*, 7, 51–63.

Fernandex, J. M., M. A. E. Selma (2004). The dynamics of water scarcity on irrigated landscapes: Mazarron and Aguilas in South-eastern Spain. *System Dynamics Review*, 20(2), 117–137.

Fletcher, E. J. (1998). The use of system dynamics as a decision support tool for the management of surface water resources. First International Conference on New Information Technologies for Decision Making in Civil Engineering Montreal, Canada, 909–920.

Ford, A. (1999). *Modeling the Environment. An Introduction to System Dynamics Modeling of Environmental Systems.* Island Press, Washington, DC.

Forester, J. W. (1961). *Industrial Dynamics.* The MIT Press, Cambridge, MA.

Forrester, J. W. (1961). *Industrial Dynamics.* The MIT Press, Cambridge, MA.

Hodges, J. S. (1987). Uncertainty, policy analysis, and statistics (with discussion). *Journal of American Statistical Association*, 2, 259–291.

Huang, G. H., N. B. Chang (2003). Perspectives of environmental informatics and systems analysis. *Journal of Environmental Informatics*, 1, 1–6.

Huggett, R. (1993). *Modeling the Human Impact on Nature*. Oxford University Press, New York.

Jian, X., Y. S. Yu (1998). A comparative study of linear and nonlinear time series models for water quality. *Journal of American Water Resources Association*, 34(3), 651–659.

Keating, E. K. (2020). Everything you ever wanted to know about how to develop a system dynamics model but were afraid to ask. https://proceedings.systemdynamics.org/1998/PROCEED/00024.PDF (accessed June 2020).

Keyes, A. M., R. N. Palmer (1993). The role of object-oriented simulation models in the drought preparedness studies. *Proceedings of the 20th Annual National Conference, Water Resources Planning and Management Division of ASCE*, 479–482, Seattle, WA.

Levin, S. A. (1985). Scale and predictability in ecological modeling. *Modeling and Management of Resources Under Uncertainty*, T. L. Vincent, Y. Cohen, W. J. Grantham, G. P. Kirkwood, J. M. Skowronski. *Lecture Notes in Biomathematics*, 72(1–8), Springer-Verlag, Berlin.

Li, L., S. P. Simonovic (2002). System dynamics model for predicting floods from snowmelt in North American Prairie watersheds. *Hydrological Processes Journal*, 16, 2645–2666.

Li, Y., C. Tang, L. Xu, S. Ye (2017). Using system dynamic model and neural network model to analyse water scarcity in Sudan. *IOP Conference Series: Earth and Environmental Science*; IOP, Bristol.

Lund, J. R., I. Ferreira. (1996). Operating rule optimization for Missouri river reservoir system. Journal of Water Resources Planning and Management, ASCE, 122(4), 287–295.

Madani, K., M. A. Mariño (2009). System dynamics analysis for managing Iran's Zayandeh-Rud river basin. *Water Resources Management*, 23, 2163–2187.

Mashaly, A. F., A. G. Fernald (2020). Identifying capabilities and potentials of system dynamics in hydrology and water resources as a promising modeling approach for water management. *Water*, 12, 1432. doi:10.3390/w12051432.

Moffatt, I. (1991). *Causal and Simulation Modeling Using System Dynamics, Concepts and Techniques in Modern Geography*. CATMOG, Australia.

Nandalal, K. D. W., S. P. Simonovic (2003). Resolving conflicts in water sharing: A systemic approach. *Water Resources Research*, 39(12), 1362–1373.

Nirmalakhandan, N. (2002). *Modeling Tools for Environmental Engineers and Scientists*. CRC Press, Boca Raton.

NRC (2001). *Assessing the TMDL Approach to Water Quality Management*. National Academy Press, Washington, DC.

Qin, H.-P., Q. Su, S.-T. Khu (2011). An integrated model for water management in a rapidly urbanizing catchment. *Environmental Modeling and Software*, 26, 1502–1514.

Reckhow, K. H., S. C. Chapra (1983). *Engineering Approaches for Lake Management, Vol. 1: Data Analysis and Empirical Modeling*. Butterworth, Woburn, MA.

Saysel, A. K., Y. Barlas, O. Yenigun. (2002). Environmental sustainability in an agricultural development project: a system dynamics approach. *Journal of Environmental Management*, 64, 247–260.

Sehlke, G., J. Jacobson (2005). System dynamics modeling of transboundary systems: The Bear River Basin model. *Groundwater*, 43(5), 722–730.

Simonovic, S. P. (2000). Tools for water management: One view of the future. *Water International*, 25, 76–88.

Simonovic, S. P. (2002). World water dynamics: global modeling of water resources. *Journal of Environmental Management*, 66(3), 249–267.

Simonovic, S. P. (2008). *Managing Water Resources: Methods and Tools for a Systems Approach*. Earthscan, Routledge, New York.

Simonovic, S. P., S. Ahmad (2005). Computer-based model for flood evacuation emergency planning. *Natural Hazards*, 34(1), 25–51.

Simonovic, S. P., H. Fahmy (1999). A new modeling approach for water resources policy analysis. *Water Resources Research*, 35(1), 295–304.

Simonovic, S. P., H. Fahmy, A. El-Shorbagy (1997). The use of object-oriented modeling for water resources planning in Egypt. *Water Resources Management*. 11, 243–261.

Simonovic, S. P., V. Rajasekaram (2004). Integrated analysis of Canada's water resources: A system dynamics model. *Canadian Water Resources Journal*, 29(4), 223–250.

Stave, K. A. (2003). A system dynamics model to facilitate public understanding of water management options in Las Vegas, Nevada. *Journal of Environmental Management*, 67, 303–313.

Sun, Y., N. Liu, J. Shang, J. J. Zhang (2017). Sustainable utilization of water resources in China: A system dynamics model. *Journal of Cleaner Production*, 142, 613–625.

Sušnik, J., L. S. Vamvakeridou-Lyroudia, D. A. Savić, Z. Kapelan (2012). Integrated system dynamics modelling for water scarcity assessment: Case study of the Kairouan region. *Science of Total Environment*, 440, 290–306.

Teegavarapu, R. S. V. (2013). Climate change-sensitive hydrologic design under uncertain future precipitation extremes. *Water Resources Research*, 49(11), 7804–7814.

Teegavarapu, R. S. V., S. Chinatalapudi (2018). Incorporating influences of shallow groundwater conditions in curve number-based runoff estimation methods. *Water Resources Management*, 32, 4313–4327.

Teegavarapu, R. S. V., A. Elshorbagy, L. Ormsbee (2002). Characterizing pollutant loadings in streams using system dynamics simulation. *Proceedings of AWRA Annual Conference*, November, 247, Welty Claire, Ed. TPS-02-4, Middleburg, Virginia.

Teegavarapu, R. S. V., S. P. Simonovic (2000a). System dynamics simulation model for operation of multiple reservoirs. CD ROM Proceedings of World Water Congress, Melbourne, Australia, 11–17 March.

Teegavarapu, R. S. V., S. P. Simonovic (2000b). Short-term operation model for coupled hydropower plants. Journal of Water Resources Planning and Management, ASCE, 126(2), 98–106.

Teegavarapu, R. S. V., S. P. Simonovic (2014). Dynamics of hydropower system operations. *Water Resources Management*. Springer Publications. 28, 1937–1958. doi:10.1007/s11269-014-0586-2.

Teegavarapu, R. S. V., A. K. Tangirala, L. Ormsbee (2005). Modeling water quality management alternatives for a nutrient impaired stream using system dynamics and simulation. *Journal of Environmental Informatics*, 5(2), 73–81.

Winz, I., G. S. Brierley, S. Trowsdale (2009). The use of system dynamics simulation in water resources management. *Water Resources Management*, 23, 1301–1323.

Wurbs, R. (1993). Reservoir-system simulation and optimization models. Journal of Water Resources Planning and Management, ASCE, 116(1), 52–70.

Xu, Z. X., K. Takeuchi, H. Ishidaira, X. W. Zwang (2002). Sustainability analysis for Yellow River water resources using the system dynamics approach. *Water Resources Management*, 16, 239–261.

Yang, C.-C., L.-C. Chang, C.-C. Ho (2008). Application of system dynamics with impact analysis to solve the problem of water shortages in Taiwan. *Water Resources Management*, 22, 1561–1577.

Chapter 3

System Dynamics Models

Applications

Ramesh S. V. Teegavarapu

3.1 INTRODUCTION

This chapter provides two case studies related to the applications of SD models for water and environmental systems. The case studies adopted from published works of Teegavarapu and Simonovic (2014) and Teegavarapu et al. (2005) are expected to help understand the SD model development process and evaluate results obtained for policy changes.

The first case study deals with a simulation model developed for modeling the dynamics of a hydraulically coupled hydropower reservoir system operation (Wood and Wollenberg, 1984). The hydraulic coupling feature is modeled using conditional constraints incorporated via functional relationships linking tailwater elevations and forebay elevations of upstream and downstream reservoirs. The incorporation of hydraulic coupling in the system results in a conceptually accurate representation of the physical system under consideration. Flow delays have a considerable impact on the power generation amounts in run-of-the-river hydropower plants with very little storage variations. The motivation to develop a simulation model is to understand (1) the behavior of the system for short-term changes in the inputs and (2) the heuristic operating procedures or rules of thumb and their effect on the overall system performance. Hydraulic and hydrologic coupling were addressed in a simulation model meant for real-time operation.

The SD model from the first case study is different in comparison with those developed in earlier works (Simonovic et al., 1997; Simonovic and Fahmy, 1999) in terms of addressing the issue of real-time operation of multiple reservoir systems considering hydrologic and hydraulic coupling and exhaustive modeling of the system and the use of two performance indices (Hashimoto et al., 1982) to quantify the system performance. The reliability and vulnerability indices are defined to assess the system status to different inflow conditions. Monte Carlo simulation is used to assess the variability of power generation due to changing system conditions.

The second case study deals with the development of an SD model for characterizing the fate and transport of a nutrient (i.e., total phosphorus) in streams. The simulation model is also used to develop a water quality management strategy for improving the health (i.e., water quality) of the stream. The model can help answer several management and policy questions and assess the vulnerability of impaired streams. One of the major advantages of this simulation model output is the generation of continuous pollutant loadings over time. Devising appropriate criteria that make use of results from simulation can help in the assessment of stream health. Two evaluation indices for the assessment of stream health are used in this study. The vulnerability index helps to evaluate how often the stream is impaired or not meeting the water quality standard. The resiliency index helps to assess the ability of the stream to recover to acceptable water quality standards after a violation of the water quality standard has occurred. These indices can help in prioritizing the impaired streams and ultimately for the development and adaptive pollution reduction strategies.

3.2 CASE STUDY I: MODELING HYDROPOWER RESERVOIR SYSTEM OPERATIONS

The reservoir system considered is a series of four hydropower reservoirs in Manitoba, Canada. A schematic representation of a two-reservoir system is provided in Figure 3.1 that is representative of the multiple reservoir systems handled. The reservoirs are linked both hydrologically and hydraulically. The former link indicates that the release from an upstream reservoir becomes an input to the immediate downstream reservoir, while the latter link suggests that any decision taken at one reservoir has an effect on the other in a way that influences the performance of both reservoirs. The link and the nature of the system are explained using a causal loop diagram. Details of such a diagram are discussed in the next section.

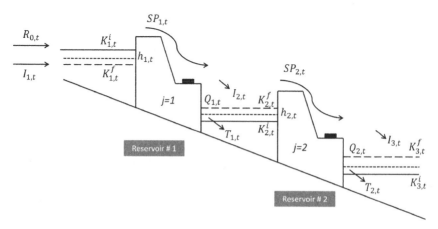

Figure 3.1 Schematic representation of a two-hydropower reservoir system.

3.2.1 Causal Loop Diagram (CLD)

A causal loop diagram following the notation suggested by Roberts et al. (1983) for a hydraulically and hydrologically coupled two-reservoir system is shown in Figure 3.2. Relationships between different elements of the system are also shown. For example, an increase in inflow would lead to an increase in storage. Similarly, an increase in tailwater elevation would lead to a decrease in the head required for power generation. The word "head" is defined as the difference in the level between the forebay and tailwater elevations of the reservoir. The variables referred to as spills, releases, reservoir levels, and heads, as shown in Figure 3.2, are defined for different reservoirs using different superscripts attached. A similar diagram can be used to explain the behavior of the next two reservoirs in a cascade and any other lakes.

Few observations can be made based on the causal loop diagram. Shocks (sudden variations) can be introduced by the exogenous inputs to any system. These inputs are the inflows (streamflows) that are both controlled and natural in the case of the hydropower system. While the first input is known a priori, the second one is stochastic. Feedbacks are evident in the system, and they are balancing or negative feedback. The balancing is due to controlled (release decisions) and uncontrolled (spill) actions in the system. For example, a rise in the reservoir level due to higher inflow affects the release made. However, as the release increases the tailwater elevation increases and thus reducing the head required for

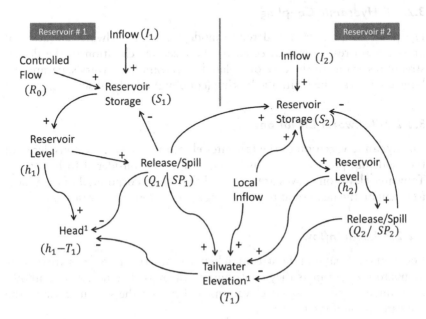

Figure 3.2 Causal loop diagram for a two-reservoir system.

energy generation. If the release is restricted, the uncontrolled spill will produce the same effect. Positive feedback behaviors are uncommon in engineering systems where human intervention mechanisms always control the dynamics of the system. The model can help plant operators to simulate conditions that occur at different time scales and understand the dynamics of power generation when there is water availability, hydraulic coupling influencing energy generation head, and operation constraints in maintaining a specific level (stage) of water in any given reservoir within a system of reservoirs.

3.2.2 Structure of the Model

The object structure of the model here refers to a series of objects (stocks and flows) with specific properties that are used to describe the model of a real system. The model developed is for a four-reservoir system in the Province of Manitoba, Canada. Different components of the model developed are specific to this system. However, the representation is generic and is easily transferable to any other system of similar configuration. A brief description of the system processes that need to be represented in the model is provided in the next several sub-sections. Mathematical equations representing different processes and functional relationships and case study specific parameters are adopted from Teegavarapu and Simonovic (2014).

3.2.2.1 Hydraulic Coupling

Hydraulic coupling is considered in modeling as the tailwater elevation of an upstream reservoir is affected by the reservoir elevation of the downstream reservoir. In the case of isolated reservoirs (reservoirs separated by large distances), the hydraulic linking (coupling) may be absent.

3.2.2.2 Tailwater Elevations

The dynamic variation of the tailwater elevation needs to be considered in the modeling process as the hydraulic linking is considered to be strong. Tailwater elevation curves are used and they are incorporated in the model to calculate the head required for energy generation at any reservoir.

3.2.2.3 Local Inflow

Local inflows that join the stream in between any two reservoirs are used as inputs to the system as they contribute to the mass balance. These inflows may influence the tailwater elevations if they join the stream close to the tailwater pool of the reservoir.

3.2.2.4 Boundaries of the System

The reservoir system considered is a set of four reservoirs. The first reservoir receives a controlled flow from the upstream reservoir that can be regarded as a source. The last reservoir in the system discharges into a lake (sink) that is considered to be the boundary of the system. The origins of local inflows into the reservoirs are considered to be the boundaries of the system.

3.2.2.5 Governing Equations

The operations of the system can be expressed as a set of equations that can help capture the dynamics of the operation process. The object structure (with stocks, flows, converters, and connectors) can be used to model these equations in the simulation environment (e.g., STELLA (ISEES, 2020)). Variation of water elevation levels within a time interval is difficult to model. Hence a relationship to obtain average elevation is used.

$$h_{j+1,t} = \left(k_{j+1,t}^i + k_{j+1,t}^f\right) 0.5 \quad j = 1, n \ \& t = 1, T \tag{3.1}$$

The above equation is used for obtaining the average forebay elevation at each of the generating stations. An average value is used as the forebay elevation fluctuates within the period considered. The variables $k_{j,t}^i$ and $k_{j,t}^f$ are the initial (i) and final (f) forebay elevations for the time interval, t, associated with the station, j, respectively.

$$h_{j+1,t} = \left(k_{j+1,t}^i + k_{j+1,t}^f\right) 0.5 \quad j = 1, n \ \& t = 1, T \tag{3.2}$$

The value of $h_{j+1,t}$ refers to the average forebay elevation of the reservoir, $j+1$. The subscript, j, t, is used instead of $j+1$, t, as $h_{j+1,t}$ is used for calculation of head at the generating station, j. For any generating station, tailwater elevation curves represent discharge-elevation curves for different downstream forebay elevations. A schematic diagram showing these elevation curves is provided in Figure 3.3.

A general form of these curves is given by the equation,

$$T_{j,t} = k_{l,j+1}^o + C_{l,,j+1}G_j^t \quad \forall j, \ \forall l, \ t = 1, T \tag{3.3}$$

where $T_{j,t}$ is the tailwater elevation and $k_{l,j+1}^o$ is a discrete downstream reservoir's forebay elevation, which is taken as a downstream condition for deriving the linear curves. The variable G_j^t represents the sum of plant discharge and spill from plant j, while the variable $C_{l,,j+1}$ is a constant in the linear expression. If the local inflow influences the tailwater elevation, it has to be included in calculating the total discharge (or project

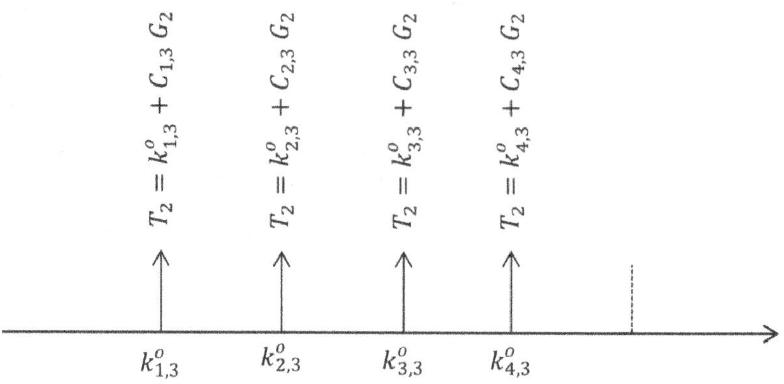

Figure 3.3 Tailwater elevation curves (functional relationships).

discharge), G_j^t. A set of tailwater elevation curves are used for different forebay elevations.

$$\beta_{j,t} = f\left(Q_{j,t}\right) \quad \forall j, \quad t = 1, T \tag{3.4}$$

The above equation defines the relationship between efficiency, $\beta_{j,t}$, and the discharge or release, $Q_{j,t}$. This constraint specifies a functional relationship between the plant discharge, $Q_{j,t}$, and the overall plant efficiency, $\beta_{j,t}$, for generating station, j, for time interval t. Efficiency, in general, is also a function of the head. In the simulation, a nonlinear functional relationship can be easily used by the inclusion of a relationship that relates efficiency with both discharge and head and poses no conceptual difficulty. However, a similar relationship defined by Equation 3.4, adopted in a weekly scheduling model, Energy Management and Maintenance Analysis (EMMA) model (Barrit-Flatt and Cormie, 1988), is used in this study.

The relationship to calculate the power generated at any hydropower generating reservoir is given by:

$$\varepsilon_{j,t} = \gamma_o \left(h_{j,t} - T_{j,t}\right) Q_{j,t} \beta_{j,t} \quad \forall j, \quad \forall t \tag{3.5}$$

where γ_o is a constant and the term $\left(h_{j,t} - T_{j,t}\right)$ provides the head for power, $\varepsilon_{j,t}$, in any time interval, at any reservoir, j, calculations.

$$S_j^{t+1} = S_j^t + I_j^t + R_{j-1,t} + R_{j,t} \quad \forall j, \quad \forall t \tag{3.6}$$

$$R_j^t = SP_{j,t} + Q_{j,t} \quad \forall j, \quad \forall t \tag{3.7}$$

$$S_j^{t+1} = S_j^t + I_{j,t} + R_{j-1,t} - O_{j,t} \quad j = n + 1, \forall t \tag{3.8}$$

The variables S_j^t and S_j^{t+1} represent reservoir storages at the beginning of time intervals t and $t+1$ in volume units, respectively. The variables $R_{j,t}$ and $SP_{j,t}$ are the release and the spill values from reservoir j, respectively, while the forecasted value of local inflow. Equation 3.8 represents the continuity equation for the lake into which the last plant is discharging water. The variable $O_{j,t}$ represents the controlled flow out of the lake.

3.2.2.6 Tailwater Curves

The hydraulic coupling that exists between any two hydropower is considered while calculating the head required for energy generation at each of the hydropower plants. The selection of the tailwater curve is based on a conditional constraint (Equation 3.9) related to the forebay elevations of the downstream reservoirs and is modeled using an if-then-else construct in the simulation environment.

$$ if\,(h_{j+1,t} > k_{l,j+1}^o \,\cap\, h_{j+1,t} < k_{l+1,j+1}^o \;\; then \;\; T_{j,t} = k_{l,j+1}^o + C_{l,,j+1}G_{j,t}) \quad \forall j,l,t \quad (3.9) $$

3.3 MODELING ENVIRONMENT

STELLA (ISEES, 2020) modeling environment is used to model the operation of a multiple reservoir system. Functional relationships and dependencies are defined using converters and connectors, respectively. Few built-in mathematical, logical, and statistical functions are used in the objects. In this case study, functions such as random, if-then-else, and delay are used. The environment-specific features such as sensitivity analysis, graphical inputs, and others are used as well. The model structure is aimed at reflecting the operation of a system of reservoirs in series, with the final reservoir discharging into a lake with a strong hydraulic coupling existing between the power plants. Three major sectors (i.e., system configuration, hydraulic coupling, and decision process) and five minor sectors for reservoirs and lakes are developed in the modeling environment. A brief description of the model sectors is given next.

3.4 REPRESENTATION OF RESERVOIRS

Each reservoir is represented as a model sector that has reservoir-specific properties. The structure of one of the main sectors (reservoir) is given in Figure 3.4. Similar structures are developed for all the other reservoirs. The reservoir-specific characteristics for a sector include stage-storage relationships, local inflow, controlled flows from any other reservoirs, discharge-efficiency

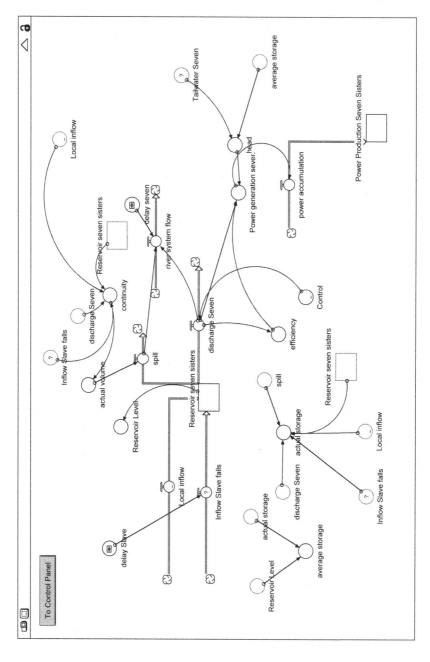

Figure 3.4 Representation of the reservoir as one of the sectors.

curves, spill calculations, plant release, average storage level, flow transport delay, and calculation of head required for power generation.

In optimization and simulation models developed in the past for hydraulically coupled reservoirs, similar rules were used and they are referred to as look-up tables. Similar rules exist for all the reservoirs that are considered. If the hydraulic coupling is neglected, the average storage value can be used for head calculation. This eliminates all the converters that house the if-then-else rules.

3.5 REPRESENTATION OF THE LAKE SYSTEM

The object structure of a lake (Winnipeg Lake) includes a stock as a water body into which the final reservoir (Pine Falls reservoir) is discharging, and also there is a controlled outflow from the lake. Converters are used to calculate the average volume of the lake and for the determination of tailwater elevation curves. Apart from these major sectors, few others are used to incorporate different features in the model. These include the selection process for variable values, graphical relationships, and all other calculations that are not included in the major sectors. A link exists between different sectors if and only if the variables in the sectors are interdependent. This can be seen in the main mapping layer of the STELLA modeling environment.

3.6 TAILWATER CURVE SELECTION

This sector uses objects with logical functions (e.g., if-then-else) to determine the appropriate tailwater elevation curve for use in the calculation of the head. A sample set of rules using the if-then-else construct using information from two reservoirs is given below.

```
IF (Average water level at reservoir B >= 246.58) AND
(average water level at reservoir B <246.74) THEN
(246.58 + (Total flow at reservoir A*0.067)) ELSE
IF (Average water level at reservoir B >= 246.74) AND
(Average water level at reservoir B < 246.89) THEN
(246.74 + (Total flow at reservoir A*0.064)) ELSE
Endif
```

3.7 MODEL APPLICATION

The real-time operation model is applied to evaluate the operation of a series of four reservoirs on the Winnipeg River in the Manitoba province in Canada. These four reservoirs in cascade form a part of a much more complex network of hydropower reservoirs maintained by the local hydropower corporation, Manitoba Hydro. Only the plants on the Winnipeg River maintained by Manitoba Hydro are used as a testbed for the developed model. A schematic representation of the plants in the case study region is given in Figure 3.5. The first plant, Seven Sisters, receives the controlled flow from the Slave Falls reservoir located upstream, while the last plant under consideration, the Pine Falls reservoir, drains into Lake Winnipeg. The McArthur reservoir has the largest storage of all the plants. Strong hydraulic coupling between the hydro-generating plants on the Winnipeg River is one of the important features of the present system. More details of the case study region are available from earlier works by Teegavarapu and Simonovic (2000) and Barritt-Flatt and Cormie (1988). The weekly target power production for all the power plants, initial and final forebay levels, and the forecasted values of inflows are assumed to be known. To facilitate development, the model is divided into three major sectors (i.e., system configuration, hydraulic coupling, and decision process). The major

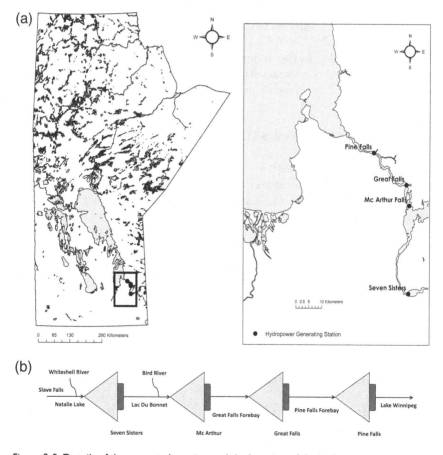

Figure 3.5 Details of the case study region and the location of the hydropower reservoirs.

sectors and the initial interactive layer of the STELLA model are shown in Figures 3.6 and 3.7. One of the advantages of dividing the model into sectors is that each sector can be run individually or multiple sectors can be run as a group. The sectors also help incorporate the required modularity in the model structure.

The SD model is validated by three tests that include: (1) replication, (2) sensitivity, and (3) prediction. These tests will confirm the structure of the model with the physical system that is modeled. The replication test is carried out by using real-life data and also a simulation experiment to compare the results obtained using release decisions by an already existing optimization model (Teegavarapu and Simonovic, 2000). The simulation model

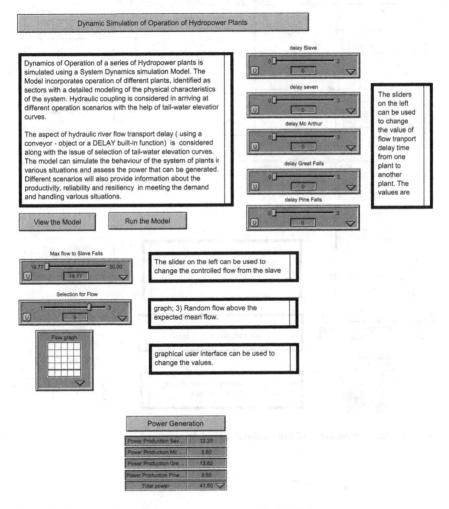

Figure 3.6 Major sectors of the model represented in STELLA.

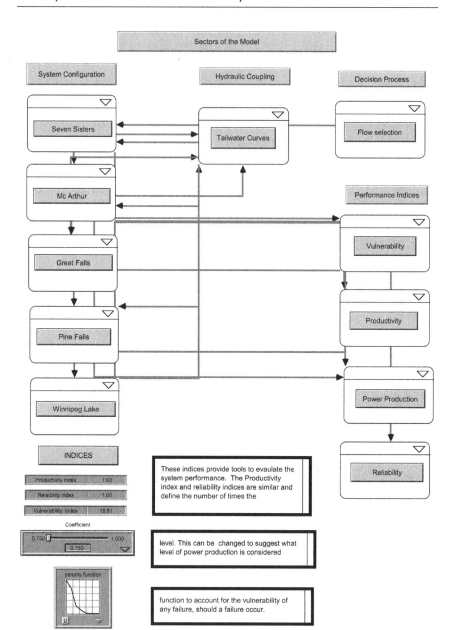

Figure 3.7 Interactive layer of the model in STELLA.

developed is not dependent or based on the optimization model. However, the optimal releases obtained from the model were used in the simulation model to evaluate the system performance with or without changes to these releases. The optimization model (Teegavarapu and Simonovic, 2000) was originally developed in the General Algebraic Modeling System (GAMS) modeling language (Brooke et al., 2006). The original optimization model development required a specialized language referred to as algebraic modeling language (AML) and a GAMS language-based syntax. The model is executed using a mixed-integer nonlinear programming (MINLP) solver (DICOPT (DIscrete Continuous OPTimizer)), and the formulation required binary variables, which resulted in computational intractability and execution time issues relevant to the real-term operation of the same coupled hydropower system. The optimization model (Teegavarapu and Simonovic, 2000) was used for short-term optimal operation and not applicable for real-time operation due to computational issues.

The model reported uses SD principles for simulation of a hydrologically and hydraulically coupled hydropower system. The main advantage of this model is that it can be used for simulation in real-time, scenario generation, and performance evaluation of hydropower energy generation systems. The simulation model is tested by using the optimal release decisions of the optimization model to replicate the storage variations as obtained by the previous model. This test can also be used to validate the prediction aspect of the system. The sensitivity of the model to different inputs is tested by using extreme events (high and low inflows). Various other tests that validate the dimensional consistency of the equations used and the robustness of the model in handling extreme conditions are carried out before the model is adopted for implementation. The model is validated using all these tests. Different initial storage conditions and the value of controlled flow into the system are considered for performing the sensitivity analysis. More scenarios can also be obtained by changing the flow transport delay between the plants. The plant discharge at each reservoir can be varied either graphically or based on any existing operating policy provided by an optimization model. This will result in different power generation values. The model can be run to obtain the operation rules using different initial values of initial storage states while retaining a constant inflow scheme to the reservoir system. The plant discharge values that are used are optimal and are obtained from an optimization study by Teegavarapu and Simonovic (2000). However, these values are not always implemented in real time and need to be refined. Sensitivity analysis can be easily performed using a simulation environment to generate a variety of operation rules.

The flow transport time between the plants can be used to evaluate the effect of the delay time on the total amount of power produced at each of the plants. The model developed in this study uses a fixed delay approach. A built-in function, delay, or an object, conveyor (in STELLA), can be used to introduce the flow transport delay. A conveyor as a stock object has special properties that allow it to convey material (in this case flow) at a pre-defined time interval. The modeling environment can also be used to incorporate a routing model. Several stocks can be used in between the two reservoirs to model the storage effect of the stream. A fixed delay of one day is used to understand its effect on power production at each of the plants. In general, flow transport delay reduces the amount of power that can be generated within the operating horizon, as some amount of water is not available to provide an adequate head for power generation. The total power production values without and with flow delay are different and a reduction in power generation will be noted due to flow delays. The power generation values will change based on the location where the delay is introduced.

3.7.1 Influence of Hydraulic Coupling

Hydraulic coupling between any two hydropower reservoirs is addressed using tailwater elevation curves which are conditioned on the downstream forebay elevations. A total of 5, 11, 11, and 15 tailwater curves are used for each of the reservoirs in the order from the upstream reservoir to the lake. When no coupling exists, the energy generation head is calculated using the average forebay and tailwater elevations or using constant tailwater elevations. The total power production values with no hydraulic coupling and without and with flow delay of one day are equal to 46.40 and 40.87 GWhrs, respectively (Teegavarapu and Simonovic, 2014). In a hydropower system where strong hydraulic coupling exists, the assumption of no coupling has resulted in increased power production values due to no influence of tailwater elevation variations on available head calculations. These energy values are 3.5% and 5.5% higher than those from similar conditions with coupling. However, the reduction of total energy value in the scenario of no coupling is 8.82% when a delay is considered, which is lower than the case when the coupling is incorporated in the model. The energy production values have increased at each of the plants when no hydraulic coupling was considered. The increase in energy production value is higher for the second plant (McArthur) compared to that of others when a delay was considered in case of no hydraulic coupling. The McArthur plant with the highest storage has the advantage of storing the delayed flows to improve the power production from a no-delay condition.

3.7.2 Operation Rules

Reservoir operating rules are often refined and re-defined based on a variety of situations in real time. The judgment and experience of reservoir operators become useful in many situations. One such case where heuristics are used would be to increase the release whenever the forebay elevation of a reservoir is increasing to avoid a spill. However, the reservoir operator's objective is to release as much water as possible to maximize the hydro-power generation that may influence the tailwater elevation which affects the head. Incorporation of different heuristic rules at different reservoirs would lead to a system behavior that is not easily predictable or counter-intuitive. If hydraulic linking is negligible, the tailwater rise or variation will not be observed. In the case of extreme events (high or low inflows), the operating rule would depend on the energy demand and storage available.

3.7.3 Performance Measures

Indices such as reliability and vulnerability (Hashimoto et al., 1982) are used to measure system performance. The reliability of the system can be assessed by the number of times the target demand is met during the whole operating horizon. Vulnerability is defined in terms of a monetary value attached to a particular failure decided by a penalty function provided by the user. These penalty functions are similar to membership functions (Teegavarapu and Elshorbagy, 2005), where the decision-maker's prefer-ences are attached to specific performance measures, which are modeled as linear, nonlinear, or sigmoid curves with penalty value confined to a specific range. These two indices are appropriate for hydropower systems as the performance of the system will be measured based on the number of times the target electric power demand is met and the specific time interval in which the target demand is not met. Vulnerability identifies the severity of failure in case a failure occurs. Loss functions are developed to obtain a penalty value (monetary) associated with a particular failure (e.g., fail-ure to meet the target demand). These curves can be modified by the user with the help of a graphical user interface (GUI) provided by the simulation environment.

The reliability and vulnerability of the system largely depend on inflows that form the major inputs to the system. Variants of the reliability index are possible and are associated with the number of times the pre-specified demand is met. Two reliability indices (λ and λ^o) developed are described by equations 3.11 and 3.12.

$$\phi = \sum_{t=1}^{T}\sum_{j=1}^{n}\varepsilon_{j,t} \tag{3.10}$$

$$if\,(\phi > \phi_{\text{Target}}, \text{ then } \lambda = 1, \text{ else } \lambda = 0) \tag{3.11}$$

The variable ϕ is the total energy produced during the entire time horizon of operation given by the summation of energy produced at each reservoir and in all time intervals within the total time horizon, and λ is the reliability of the system, and ϕ_{Target} is the target energy from the system. Alternatively, a variant of reliability index (λ) can be expressed as a ratio of the number of time intervals the energy production exceeded the target to the total number of time intervals (months). The refined definition of reliability, λ^o, is given by:

$$\lambda^o = \frac{\nu}{T} \tag{3.12}$$

where ν is the number of times the energy production exceeded the target and T is the total number of time intervals in the operating time horizon under consideration. The vulnerability level of achievement ratio is defined based on the total energy value (ϕ) produced in comparison to the target value ϕ_{Target}. The vulnerability index indicates the cost/penalty associated with a failure. The vulnerability index (φ) is calculated based on a performance function $f(\psi)$, defined based on a total deficit.

$$\psi = \frac{\phi}{\phi_{Target}} \tag{3.13}$$

$$\varphi = f(\psi) \tag{3.14}$$

A penalty function similar to a membership function generally defined in the fuzzy set theory can be developed for Equation 3.14. Estimates of the reliability and vulnerability for this case study system are evaluated and discussed by Teegavarapu and Simonovic (2014) and are not presented in this chapter. Smaller values of the index suggest the better performance of the system. It should be noted that there is no conceptual difficulty in developing these criteria for the whole system. One of the major advantages of the SD approach is that qualitative information can be handled. For example, if two operating policies are available, then the operator's willingness to use any one of the policies can be incorporated through graphical functions within the simulation environment. The sensitivity analysis feature of the simulation environment is particularly useful for reservoir/plant managers to generate operating rules for various uncertain conditions. This feature is also useful in instances where the time interval within which decisions are required is too small to run an optimization model. The model can be run to obtain the operation rules using different values of initial storage states while retaining a constant inflow scheme to the reservoir system. The model can be extended to include the simulation of individual turbine operations at each of the reservoirs. To achieve a complete representation of the physical system (e.g., incorporation of hydraulic coupling and calculation of average storage elevations), additional objects are defined in the model.

The use of if-then-else rules for the selection of tailwater curves introduces a form of rigidity into the system that influences the dynamics of the system. The rigid structure of these rule constructs may lead to discontinuity in some instances, especially where the variation in system conditions is difficult to model. The dynamic variation of tailwater elevation due to changes in forebay elevation can only be modeled for discrete intervals of these elevations as the tailwater elevation curves are available for these discrete water levels. Any behavior of the system within the intervals can be associated with the use of one of the tailwater elevation curves. The system representation can be improved if the number of tailwater elevation curves is increased. In summary, the model uses objects with specific properties that provide an accurate representation of the physical system under consideration. The simulation environment provides an easy mechanism to include physical aspects of the system which otherwise are difficult to incorporate into traditional simulation models. The scenarios generated will be useful for the real-time implementation of the operating rules. Even though the model is specific to reservoir operation related to hydropower generation, the modeling concepts can be extended to any type of reservoir system with any type of objective.

3.7.4 Sensitivity Analysis: Monte Carlo Simulation

The object-oriented simulation environment provides a sensitivity analysis feature to generate multiple runs of the simulation model by perturbing a single parameter or variable value in the model. The feature includes the advantage of generating the value based on a probability distribution. Uniform and normal distributions are allowed in the STELLA modeling environment. By using the distribution of the inputs, the sensitivity analysis process, therefore, becomes a Monte Carlo simulation exercise. In the system under consideration, an important variable that affects the operations and power generation is the flow to the first reservoir (Seven Sisters Reservoir) in the cascading system of reservoirs. This flow is important as the system has little storage at all the reservoirs. Monte Carlo simulation is conducted by perturbing this flow variable by assuming a uniform distribution of values within the range of $[0.504-0.586]\times10^3$ m³/sec. The random number generator available in the simulation environment is used to generate 1000 of these flow values. The distributions of flows and total power generation for the entire cascading reservoirs for seven days indicate that the power generation values are skewed to right as the frequency of flow values in the range of $[0.504-0.586]\times10^3$ m³/sec are higher. Using different values of flows, it is possible to generate an ensemble of power generation scenarios and also exceedance probabilities associated with these energy levels. The latter is possible using a cumulative distribution function (CDF) based on the Monte Carlo simulation runs. The distribution of total power generation values was near normal when flow values to the first reservoir used in the system follow a Gaussian distribution. The modeling of the multiple reservoir system using the system dynamics approach has helped in understanding the dynamics of the operation. The hydropower reservoir system SD simulation model will have enormous utility in a variety of real-time operational conditions. The advantages of the model include (1) generation and evaluation of different operating rules based on a variety of conditions related to inflows, initial storage levels, and tailwater conditions; (2) evaluation of system performance through different indices; and (3) usage for real-time operational decisions. The object-oriented simulation (OOS) environment is ideal for application to reservoir operation problems. Reliability and vulnerability used as indicators to quantify the system performance in response to different inflow conditions can benefit decision-makers involved in the economic side of system operation.

3.8 CASE STUDY II: STREAM WATER QUALITY MANAGEMENT – NUTRIENTS

Nutrients such as phosphorus, nitrogen, and carbon are generally considered to be vital to sustaining aquatic ecosystems. However, an abundance of these nutrients will, in some conditions, accelerate the natural eutrophication process of a water body and is also considered to be an interference with desirable water uses (Thomann and Mueller, 1987). Eutrophication, a process stimulated by an increase in nutrients, leads to nuisance algae blooms, or more commonly periphyton (rooted algae), in swift-moving freshwaters. Anthropogenic changes to ecosystems and water bodies through wastewater treatment effluents in streams and/or agricultural practices and fertilizer uses are generally considered as causes for the imbalance of nutrients concentrations in streams and water bodies. Nutrient impairment of streams is a major problem plaguing many streams and water bodies in the U.S. and many other parts of the world. Section 303(d) of the Clean Water Act (CWA) and the United States Environmental Protection Agency's Water Quality Planning and Management Regulations require states in the U.S. to develop total maximum daily loads (TMDLs) for their water bodies that are not meeting designated uses under technology-based controls for pollution.

The TMDL development process establishes the allowable loadings of pollutants or other quantifiable parameters for a water body based on the relationship between pollution sources and in-stream water quality conditions (Lung, 2001). The TMDL process enables the states to establish water quality–based controls to reduce pollution from both point and non-point sources and restore and maintain the quality of their water resources (USEPA, 1998). To achieve this objective, the fate and transport of pollutants in streams must be modeled first to facilitate the process of assessing the source and magnitude of the relative pollution loads generated by the point and the non-point sources. This objective highlights the necessity of developing both modeling and environmental policy analysis tools. Modeling the fate and transport of nutrients (USEPA, 1999a) in water bodies under data-poor situations is a challenging problem. The main difficulty of nutrient/eutrophication modeling is due to nonlinear interactions between nutrients and plants (Thomann and Mueller, 1987). Craig et al. (2000) point to difficulties associated with the calibration of physically based models and estimation of parameters from limited data. In many situations, the parameters of water quality models cannot be uniquely obtained from the available field data and thus must be estimated from technical guidance documents (Bowie et al., 1985). In such situations, little confidence can be attached to the results of the models (NRC, 2001) and therefore they are not appropriate to evaluate future environmental management scenarios. On the other hand, simple inductive models can provide valuable insights into the processes and at the same time not being highly parameterized. Hodges (1987) and Levin (1985) indicate that overly detailed models are useless as predictive devices and suggest that techniques for aggregation and simplification are essential. The model selection criteria concerning cost, flexibility, adaptability, and ease of understanding all tend to favor conceptually simple models and promote research for the development of models that can be fully parameterized from the available data (NRC, 2001).

In this case study, the SD approach is adopted to model the spatial and temporal variability of total phosphorous (TP) in a nutrient-impaired stream in the state of Kentucky, USA. The objectives of this study are: (1) to model the transport and fate of total phosphorous through the 12 mi of impaired stream in the Town Branch watershed; (2) to develop a TMDL from the obtained model and to generate what-if scenarios; and (3) to simulate and analyze water quality strategies for this specific stream and evaluate the outcomes of different possible future scenarios in the region.

3.9 NUTRIENT MANAGEMENT MODEL

The development of a water quality model using the object-oriented simulated environment, STELLA, is briefly discussed here. More exhaustive details of the model are discussed by Teegavarapu et al. (2005). A causal loop or influence diagram (Roberts et al., 1983; Moffat, 1991; Sterman, 2001) for a river system polluted by a wastewater treatment plant (WWTP) is shown in Figure 3.8. An increase in pollutant loads from the WWTP increases the pollutant concentrations in the stream, and an increased inflow in the river increases dilution and therefore reduces the concentration levels in the stream. The elements depicting concentration of pollutant, water quality standard and vulnerability of stream in Figure 3.8 contribute toward information feedback that will help in improving the water quality of the stream by adopting pollution abatement measures (e.g., total maximum daily load implementation). The material feedback is achieved by the physical transfer of quantities between stocks through flows. Since

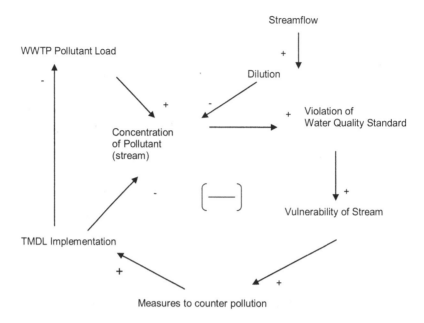

Figure 3.8 Causal loop diagram for nutrient management problem. Source: Teegavarapu et al. (2005).

pollution abatement measures will be undertaken through the implementation of a pollution load reduction strategy, in the future, the whole system will be governed by a negative feedback behavior that tries to stabilize the pollutant loads in the system. This is shown in the center of Figure 3.8 following the rules for the development of CLDs. After the creation of CLDs, stock and flow diagrams are developed and simulation is carried out.

The SD simulation model is developed to understand and model the fate and transport of a nutrient in a stream. To model the transport of total phosphorous through the stream an object-oriented simulation environment, STELLA (HPS, 2000), is used. To simplify the analysis, the stream system is characterized as a plug flow system with negligible dispersion. Assuming steady-state conditions, the spatial variability of the concentration along the river is modeled. The stream is divided into several reaches. Stocks are created at different reaches which accumulate the loads and then the flow objects pass the loads to the next reach. The loads are thus propagated through flows to the respective stocks and the obtained loads at each stock are again propagated to the next reach taking into consideration the decay represented by a factor "k." Different values of "k" are initially assumed for different reaches. The "k" factor is assumed to consider decay, deposition, chemical, and biological degradation, etc. Advection and dispersion processes are not considered explicitly in the model. However, it is assumed that these processes are implicitly incorporated into one single decay coefficient in each reach through a first-order decay equation. The total phosphorous concentration is modeled as a substance that decays as a first-order process (Thomann and Mueller, 1987) given by Equation 3.15.

$$C = C^o e^{-kt} \tag{3.15}$$

The variables C and C^o represent the total phosphorous concentrations (mg/l) at downstream and upstream reaches, respectively. "k" is the decay rate (per day) and "t" is the travel time in days that is estimated from the distance and velocity of the stream. Several assumptions are made to use the simplified concentration equation given in Equation 3.15, and these include: (1) the flow rate in the stream is independent of stream length and travel time; (2) the stream cross-section is prismatic and constant over time and length; and (3) all the reactions are of first order and are consumptive with a rate constant, k (T^{-1}). Equation 3.15 is used to predict the spatial and temporal variations of concentration in the stream system. The existence of a relationship between dissolved oxygen (DO) and the phosphorus concentration in a stream is not considered in this model. However, the structure of the model can be easily modified to incorporate such a relationship if extensive sampling results of dissolved oxygen and phosphorus concentrations are available. The boundaries of the system are represented by a source and sink objects representing the starting and ending points of the reach.

The model for two representative reaches (A and B) within a river system is shown in Figure 3.9, where the stocks represent the loads that accumulate at different reaches. The loads are propagated through "Flow" objects. Converters are used to represent the concentrations, flow time series, and to represent the first-order decay process. The connectors pass the information relevant to the exponential decay rate from one converter to the other. The object structure (Figure 3.9) also shows WWTP discharging its effluent in the river between reaches A and B.

The concentration at the confluence of WWTP and river is calculated using the combined flow and concentration values at the confluence. Dimensionality and replication tests are conducted on the basic structure of the model by checking the integrity of mass balance equations and by assessing the results of the model after it is calibrated and applied to validation data. Sensitivity tests are carried out to check the sensitivity of the model results to the parameters used in the model.

3.10 APPLICATION OF WATER QUALITY MODEL (WQM)

An SD model is developed and is used to evaluate a nutrient-impaired stream in the Town Branch watershed of Kentucky as shown in Figure 3.10.

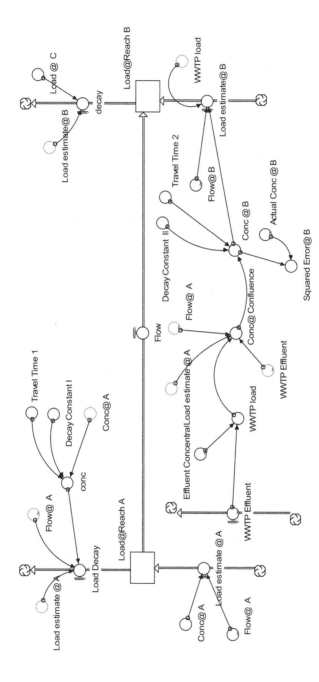

Figure 3.9 Flow and stock diagram for the modeling of fate and transport of nutrient between two locations. Source: Teegavarapu et al. (2005).

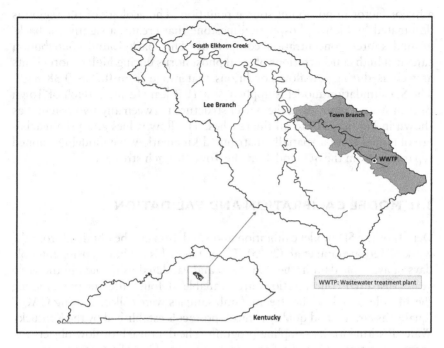

Figure 3.10 Location of the nutrient-impaired stream and wastewater treatment plant in the case study region.

The impairment is mainly caused by the high-nutrient loadings from the effluent of wastewater treatment plant discharging into the stream and also due to the discharge of nutrients from urban and agricultural run-off in the region. The Town Branch is a third-order stream that originates in downtown Lexington and flows northwest where it joins with South Elkhorn Creek at river at 54.2 km (34 mi). The Town Branch's main stem is approximately 18.5 km (11.5 mi) long and drains an area of 94.54 km² (36.5 mi²), most of which, the upper part, is an urban development in the city of Lexington. Since the effluent from the Town Branch wastewater treatment plant (WWTP) constitutes the majority of flow in Town Branch creek, phosphorus concentrations from the treatment plant dominate the total phosphorus loads in the system. The WWTP is shown as a black and white circle in Figure 3.10.

The Town Branch watershed is in the Inner Blue Grass physiographic region of Kentucky. The area is underlain by the Lexington limestone formation of the Ordovician age. Land use in the Town Branch watershed is grouped into three main categories: urban (5.5%), rural (44%), and agricultural (50.5%). The headwaters of the basin are heavily impacted by urban and suburban areas. The Town Branch watershed is unique to the nutrient-loading quantification, in the sense that background sources play a major factor in non-point source pollution. The geology of the region is dominated by highly phosphatic limestone that creates a significant background source concentration component. This background contribution can yield high concentrations of total phosphorus during high runoff events as well as during low-flow conditions that range from 0.2 to 0.38 mg/l. The SD simulation model is applied to a 19.3 km (12 mi) stretch of Town Branch stream replicating the object structure between any two reaches, as shown in Figure 3.10, to all the reaches. The flow values at all the reaches are obtained from a spatially distributed kinematic wave model developed separately from the SD model for the Town Branch stream.

3.11 MODEL CALIBRATION AND VALIDATION

Details of the SD model calibration and validation can be obtained from the work of Teegavarapu et al. (2005). The SD model is calibrated using multiple days of available data in the year 2000 and validated using data in the same year. Measured concentrations are available at four different points along the 14-mile stretch of the stream. Grab samples were collected using QAQC (quality assurance and quality control) protocols established by the Kentucky Water Institute for its sampling program. The data used for flow objects relevant to treatment plant discharge or streamflow should be continuous values in STELLA as the modeling environment uses a finite difference numerical scheme for computations. Automatic calibration using optimization procedures within STELLA is not possible as it is a simulation environment. Manual calibration to obtain the best values of decay parameters is initially carried out by trial and error method, and the range of decay parameter value for each reach is ascertained based on visual assessment of simulation results. The model is then iteratively executed by using random values of the parameter (decay parameter, k) with uniform distribution within a previously determined range, and the mean squared error (MSE) based on observed and predicted concentrations is calculated for each run. Optimal values of "k" are then selected using the lowest MSE criterion and acceptable model performance at each of the reaches. The sensitivity analysis feature of STELLA is used in this process to refine the calibration process. The observed and predicted concentration values evaluated suggest that they are in good agreement with each other (Teegavarapu et al., 2005).

To assess the effect of the decay parameters on the model performance, model calibration is carried out using two data sets. Dataset 1 is the observed data available for June in the year 2000, and data set 2 is based on the observed data available for November in the same year. The values of "k" are higher for data set 2 than the values relevant to data set 1. While the reasons for this seasonal variation of "k" values are not completely obvious. However, these results suggest that there is a need for developing two separate models for two different data sets. The performance of the model quantified by MSE improved when two data sets were used to obtain separate decay parameters and are used in the model. The model results are sensitive to the seasonal variation of decay parameter values. An improved understanding of the seasonal variation of "k" values might be possible by collecting more data for calibration and validation. More details of the calibration and validation of the SD model are provided by Teegavarapu et al. (2005).

3.12 RESULTS AND DISCUSSION

The calibrated and validated nutrient impairment SD model is used for the assessment of water quality management alternatives by modifying the loads from the point and non-point sources, the vulnerability of stream to impairment, and finally the TMDL development. Assessment of water quality is made by modifying point loads from WWTP that dominate the stream under consideration to check any spatial or temporal violations of water quality standards. Effluent concentrations of WWTP ranging from 0.5 to 3.5 mg/l with increments of 0.5 mg/l are used to determine the concentrations at the point of concern. A simulation exercise was carried out for the year 1999 and the results suggested based on the point of concern (i.e., approximately 10 mi downstream of WWTP) and water quality standard that load reductions can be estimated at the only source of interest, WWTP. The simulation model is also used to calculate the vulnerability of the stream to nutrient impairment by finding the number of days the stream is impaired if an acceptable concentration level for total phosphorus is set at any point of concern. To assess the vulnerability of the stream, the acceptable value of concentration at the point of concern is set at 0.5 mg/l. Results indicate that effluent concentration from WWTP should be equal to 0.5 mg/l to achieve the status of no impairment at the point of concern. The number of impaired days is zero when the effluent concentration is 0.5 mg/l, which points to no vulnerability.

3.13 TOTAL MAXIMUM DAILY LOAD

The total maximum daily load (TMDL) describes the maximum amount of pollutant load that a stream or a waterbody can assimilate without leading to a violation of water quality standard(s). TMDL is comprised of the sum of individual waste load allocations (WLA) for point sources, load allocations (LA) for non-point sources, and margin of safety (MOS) that accounts for uncertainty in the relationship between pollutant loads and the quality of receiving water body. The TMDL is given by Equation 3.16.

$$TMDL = \sum WLA + \sum LA + MOS \qquad (3.16)$$

In the development of TMDL for the impaired stream under consideration, load allocations that reflect non-point sources are not considered. Concentrations related to background sources (non-point sources) are already considered in the simulation model. The margin of safety is generally implicitly included in the TMDL by using conservative model assumptions to develop allocations (USEPA, 1991, 1999b). This is achieved by using conservative estimates of background concentrations. The existing wastewater treatment plant in the Town Branch watershed is considered to be the major point source contributing to the pollution at the downstream point of concern and the TMDL is developed accordingly. Load allocations are not sought for non-point sources. The total maximum daily load is obtained by limiting the WWTP effluent to permitted design discharge (i.e., 30 million gallons per day (MGD)) and estimating the load reductions for achieving required water quality standard at the point of concern. In this case study, this is achieved by first taking a critical flow year (i.e., 1999) guided by the lowest six-month average flows in the year selected from the years 1980–2000. The concentrations of nutrients along different reaches of the stream were simulated by varying the effluent concentrations of WWTP. The state of Kentucky had no official numerical (water quality) standard or criterion for total phosphorus or total nitrogen in 2005 when this model was developed. There are recommendations for allowable phosphorus concentrations to prevent nutrient over-enrichment from USEPA (United States Environmental Protection Agency). In general, any concentration of phosphorus over 0.1 mg/l has the potential to cause eutrophication in a stream. However, the trigger values of concentrations for the onset of eutrophication vary from one water body to another and therefore these values are not fixed. Numerical targets of nutrient concentrations expressed as total phosphorous (TP) are generally used to address the nutrient availability issue in streams (USEPA, 1999a). In one scenario to develop the TMDL, the allowable concentration at the point of concern is set at 0.7 mg/l and the effluent concentration from WWTP is set at a six-month average value of 2.2 mg/l. To achieve in-stream total phosphorus concentration below 0.7 mg/l at the point of concern, the Town Branch WWTP effluent concentration must be reduced from the observed six-month average of 2.2 mg/l to approximately 1.0 mg/l. In this scenario, the effluent concentration of 1 mg/l of WWTP will result in the 0.63 mg/l at the point of concern. Therefore, the allowable load should be derived based on the allowable effluent concentration and discharge of WWTP. Using maximum effluent concentrations of 2.2 and 1 mg/l from the Town Branch WWTP (with an assumed permitted daily discharge of 30 million gallons per day (MGD)) (46.4 cfs) would yield 550 and 250 lb/day of phosphorus loads, respectively. The waste load reduction for WWTP is equal to the difference of the observed six-month average load and allowable load, i.e., 300 lb/day, and the TMDL is 250 lb/day. Based on the allowable concentration limit at the point of concern and permitted discharge of WWTP and for any specific chosen critical year, the TMDL will correspondingly change. The need for TMDL development itself can also be established based on the concept of vulnerability. This is achieved by simulating the model under real-time conditions and observed data to assess the impairment.

3.14 WATER QUALITY ASSESSMENT MODEL FOR PATHOGENS

A limited number of research studies have been reported in the past to model pathogenic organisms in streams and large water bodies at a watershed scale. Modeling of fecal coliform in streams is proposed from a management perspective at the watershed level in a work by Teegavarapu et al. (2002). The fate and transport of fecal coliform in a watershed in Southeastern Kentucky, using a simulation model, based on the concepts of system dynamics (SD) approach was developed. The approach combines both data-driven approaches and insights gained from a process-based approach. Different management scenarios, based on flow conditions and pollution sources, are generated and evaluated to validate the proposed approach. Deterministic and conceptually simple probabilistic analyses are carried out to understand several water quality management alternatives that aim to reduce pollutant loadings. Results from this model point to the potential use of the proposed SD principles in addressing environmental policy issues and also to the need for relying on probabilistic analysis to obtain more credible results and recommendations in data-poor conditions. The SD model can be used to allocate limited funding for different watershed management options to be focused on areas that have the greatest impact on surface water quality conditions.

3.15 GENERAL OBSERVATIONS

SD modeling environment used to develop models for a hydraulically coupled reservoir system and a nutrient-impaired stream is discussed earlier in this chapter. Object-oriented simulation environment based on SD principles such as STELLA is often referred to as a dynamic modeling environment (Costanza and Gottlieb, 1998; Hannon and Ruth, 1997) with the ability to mainly handle and model temporal changes in a system. Spatial representation of the physical system is also possible as demonstrated by the two case studies discussed. Simulation models developed using system dynamics principles are designed to understand the basic structure and behavior of the physical system. Therefore, trends associated with variable values or alternative policy decisions derived based on the model results are oftentimes more important than the actual numerical values obtained through the simulation. Simulation models developed using any other platforms (e.g., spreadsheet-based models and others developed using formal programming languages) lack transparency while models developed using an object-oriented simulation environment with SD principles allow the modeler to evaluate the structural and behavioral aspects of the processes included in the model. The reader is directed to a recent state-of-the-art review by Mashaly and Fernald (2020) which provides a comprehensive list of developed SD models confirming the utility and applicability of SD principles to several water and environmental systems.

3.16 SUMMARY AND CONCLUSIONS

Development and applications of SD models for operation and management of water and environmental systems using two case studies are described in this chapter. A simulation model for the operation of a four-reservoir system for hydropower production and a management model for a nutrient-impaired stream are used to illustrate the utility of the SD approach. The two case study applications demonstrate the utility of SD concepts to real-world problems and complexities involved in the calibration and validation of the models. System behaviors are evaluated under different scenarios by changing the inputs. The studies presented in this chapter also demonstrate the utility of SD-based models for real-time operation of systems and policy analysis for the implementation of one of the best options derived from a multitude of scenarios for water quality management.

REFERENCES

Barrit-Flatt, P. E., A. D. Cormie (1988). A comprehensive optimization model for hydro-electric reservoir operations. *Proceedings of Computerized Decision Support Systems for Water Managers*, J. W. Labadie, et al., ASCE, 463–477.

Bowie, G. L., W. B. Mills, D. B. Porcella, C. L. Campbell, J. R. Pagenkopf, G. L. Rupp, K. M. Johnson, P. W. H. Chan, S. A. Gherini, C. E. Chamberlin (1985). Rates, constants, and kinetic formulations in surface water quality modeling. EPA/600/3-85/040, U.S. Environmental Protection Agency, Washington, DC.

Brooke, A., D. Kendrik, A. Meeraus (2006). *GAMS: A User's Guide*, 286 pages.

Costanza, R., S. Gottlieb (1998). Modeling ecological and economic systems with STELLA: Part II. *Ecological Modeling*, 112, 2–3, 81–84.

Craig, A. S., M. E. Borsuk, K. H. Reckhow (2000). Nitrogen TMDL development in the Neuse River Watershed: An imperative for adaptive management, project report. *Environmental Sciences and Policy Division*, Nicholas School of the Environment and Earth Sciences, Duke University.

Hannon, B., M. Ruth (1997). *Modeling Dynamic Biological Systems*. Springer-Verlag, New York.

Hashimoto, T., J. R. Stedinger, D. P. Loucks (1982). Reliability, resiliency and vulnerability criteria for water resource system performance evaluation. *Water Resources Research*, 18(1), 14–20.

Hodges, J. S. (1987). Uncertainty, policy analysis, and statistics (with discussion). *Journal of American Statistical Association*, 2, 259–291.

HPS, High Performance Systems (2000). *STELLA Reference Manual*. High Performance Systems Inc., New Hampshire.

ISEES (2020). IEEE systems. https://www.iseesystems.com/resources/tutorials/legacy/ (accessed October 2020)

Levin, S. A. (1985). Scale and predictability in ecological modeling. *Modeling and Management of Resources Under Uncertainty*, T. L. Vincent, Y. Cohen, W. J. Grantham, G. P. Kirkwood, and J. M. Skowronski, Springer, Berlin.

Lung, W-S. (2001). *Water Quality Modeling for Wasteload Allocations and TMDLs*. John Wiley, New York.

Mashaly, A. F., A. G. Fernald (2020). Identifying capabilities and potentials of system dynamics in hydrology and water resources as a promising modeling approach for water management. *Water*, 12, 1432. doi:10.3390/w12051432.

Moffat, I. (1991). *Causal and Simulation Modelling Using System Dynamics, Concepts and Techniques in Modern Geography*. Environmental Publications, Norwich

NRC (2001). *Assessing the TMDL Approach to Water Quality Management*. National Academy Press, Washington, DC.

Roberts, N., D. Andersen, R. Dea, M. Garet, W. Shaffer (1983). *Introduction to Computer Simulation: A System Dynamics Approach*. Productivity Press, Portland, OR.

Simonovic, S. P., H. Fahmy (1999). A new modeling approach for water resources policy analysis. *Water Resources Research*, 35(1), 295–304.

Simonovic, S. P., H. Fahmy, A. ElShorbagy (1997). The use of object-oriented modeling for water resources planning in Egypt. *Water Resources Management*, 11, 243–261.

Sterman, J. D. (2001). *Business Dynamics: Systems Thinking and Modeling for a Complex World*. McGraw Hill, New York.

Teegavarapu, R. S. V., A. Elshorbagy (2005). Fuzzy set-based error measure for hydrologic model evaluation. *Journal of Hydroinformatics*, 7, 199–208.

Teegavarapu, R. S. V., A. Elshorbagy, L. Ormsbee (2002). Characterizing pollutant loadings in streams using system dynamics simulation, extended abstract. Proceedings of AWRA Annual Conference, November, p. 247, Welty Claire, Ed. TPS-02-4, Middleburg, Virginia.

Teegavarapu, R. S. V., S. P. Simonovic (2000). Short-term operation model for coupled hydropower plants. *Journal of Water Resources Planning and Management, ASCE*, 126(2), 98–106.

Teegavarapu, R. S. V., S. P. Simonovic (2014). Dynamics of hydropower system operations. *Water Resources Management*, Springer Publications. 28, 1937–1958. doi:10.1007/s11269-014-0586-2.

Teegavarapu, R. S. V., A. K. Tangirala, L. Ormsbee (2005). Modeling water quality management alternatives for a nutrient impaired stream using system dynamics and simulation. *Journal of Environmental Informatics*, 5(2), 73–81.

Thomann, V. R., J. A. Mueller (1987). *Principles of Surface Water Quality Modeling and Control*. Harper and Row , New York..

USEPA (1991). *Guidelines for Water Quality-Based Decisions: The TMDL Process.* EPA 440/4-91-001, Washington, DC.

USEPA (1998). *National Water Quality Inventory.* Office of Water, U. S. Environmental Protection Agency(USEPA), Washington DC.

USEPA (1999a). *Protocol for Developing Nutrient TMDLs.* USEPA, EPA 841-B-99-007, Washington, DC.

USEPA (1999b). *Draft Guidance for Water Quality-Based Decisions: The TMDL Process.* EPA 841-D-99-001, Washington, DC.

Wood, A. J., B. F. Wollenberg (1984). *Power Generation, Operation, and Control.* John Wiley and Sons, New York.

Chapter 4

Simulation Models with Animation

Ramesh S. V. Teegavarapu

4.1 INTRODUCTION

In the previous chapter approaches for dynamic simulation using SD principles which mainly focused on the temporal evolution of the water and environmental systems were presented. In this chapter, simulation and animation approaches that can be used for assessment of spatial extents of inundation due to a catastrophic flooding event are discussed.

4.2 SIMULATION COMBINED WITH ANIMATION

Hydrologic and hydraulic data acquisition using monitoring networks, modeling, and simulation of extreme hydrologic events are the key components of disaster and emergency preparedness planning due to extreme flooding events. Advances in flood modeling and probability assessment, risk and mapping, and resilience concepts are discussed in multiple studies (Ashley et al., 2007; Sene, 2008; 2012). Water and emergency management agencies are responsible for the collection, validation, and archiving of the hydrologic data, modeling of extreme hydrologic events, and supporting state and federal agencies to prepare and plan for such events. There is a constant need for the development of modeling environments/animation schemes that are capable of visualizing these events to (i) enhance the understanding of possible damages to infrastructure and (ii) plan for public safety and emergency preparedness due to natural disasters (e.g., flooding) mainly caused by an extreme combination of hydro-meteorological processes. Sene (2008) indicates that flood emergency response applications should include flood maps with 3D views and animations of flood extent against a backdrop of topography and simulators that include virtual reality effects required for emergency response exercises.

Efficient assessment of damages to infrastructure can only be done if extreme flooding events can be visualized in space to evaluate the nature and extent of damage to specific objects of interest. In the current context, there is a need to evaluate the possible depth of flooding in a specific urban or rural area or inundation of hydraulic structures operated by water management officials, essential roads, and other infrastructure. Evacuation planning is a priority to emergency management officials from state and federal disaster management agencies, who are interested in assessing the extent of flooding in a region and identifying safe and reliable transportation routes that are not affected by flooding. Visualization and animation schemes that combine exhaustive details of the terrain, infrastructure, and flooding extent and provide an easy way of identifying optimal emergency management alternatives are always required. These schemes will also help the agencies and decision-makers in the dissemination of valuable emergency preparedness information to the public and provide an interactive assessment of flooding depths.

Visualization and simulation tools are increasingly used to present complex information in an intuitive way to non-specialists and for training exercises and assisting in developing emergency response plans (Sene, 2008). Several studies (Dalponte et al., 2006; Horritt and Bates, 2001; Romanowicz and Beven, 2003) have focused on the simulation of flooding events and the assessment of inundation areas and risks. Flood visualization with the help of geographical information system (GIS) tools was reported in several studies in the past decade (Bates and De Roo, 2000; Maidment et al., 2005). GIS-based systems are increasingly used to assist both with planning for emergencies and in the recovery phase (Sene, 2008, MacFarlane, 2005; Van Oosterom et al., 2005). These studies have focused mainly on representing a static map of the extent of flooding, depiction of water level rise over time, and development of animation movies based on the snapshots of the water level rises. The virtual environmental planning systems (VEPS) project is one of the leading examples of using virtual reality representation of flooding in a residential area. Geo-visualization (Pajorova et al., 2007) for emergency response applications is an active research area. Gyorfi et al. (2006) indicate that simulators for flood response applications should include four functionalities, and they include: (1) virtual reality; (2) multimedia; (3) networking; and (4) artificial intelligence. Visualization and "fly-by" animation of flooding extent, with the incorporation of satellite imagery and 3D structures of different features of the urban and rural landscape, has never been attempted before. Sophisticated modeling environments combined with 4D (four-dimensional: three-dimensional space and time) animation capabilities are extremely useful for such purposes. The current study explores the use of these modeling environments for simulation and animation of a catastrophic flooding event in a specific region of South Florida. A 3D visualization and animation environment is created in this study for this purpose.

4.3 SPATIAL AND HYDROLOGICAL DATA

Two important data sets are required for successful visualization of the results from hydrologic simulation models. These are categorized into two data types: (1) geospatial and (2) hydrologic data. The spatial data mainly comprises the terrain, aerial imagery, digital elevation model (DEM), and infrastructure data. The hydrologic data relates to the specific objectives of the study. In the current context, the data required are mainly the water levels (e.g., stages) obtained from a hydrologic simulation model (or other approximate methods) considering extreme hydro-meteorological inputs. The spatial data is generally available from state and federal agencies (e.g., United States Geological Survey (USGS)). The current study uses spatial data and water levels (i.e., flooding depth) data caused by a hypothetical extreme hurricane event from a local water management agency for a region of South Florida. More details about the data and sources can be obtained from Teegavarapu et al. (2007).

4.4 METHODOLOGY

The methodology for the visualization of flooding events depends on the successful execution of several tasks. Primary study tasks and their expected typical results and the order of accomplishment are shown in Figure 4.1. Several sub-tasks that also needed to be carried out under each of the primary tasks include: (i) identification and selection of a region; (ii) review of available geospatial and other data (GIS and AUTOCAD); (iii) review of water level data provided by hydrological simulation models; (iv) collection of data from other sources (including road network with elevation attributes); (v) assessment and evaluation of data requirements for the

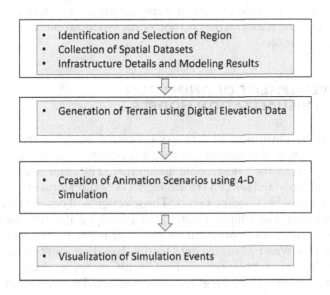

Figure 4.1 Primary steps identified for simulation and animation.

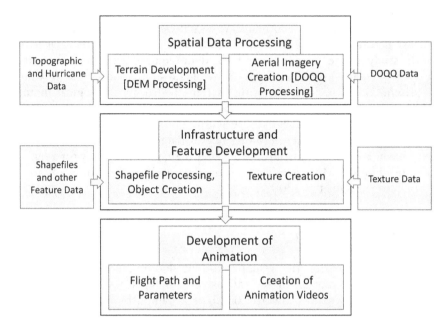

Figure 4.2 Geoprocessing of spatial data and modeling tasks.

simulation and visualization; and (vi) processing of the data for a fly-by animation and generating AVI (Audio Video Interleave) files. The methodology used in the study comprises several primary steps and are identified in Figure 4.2. The most important steps include the generation of the terrain using spatial data and the creation of the visualization environment for creating animations.

4.5 DEVELOPMENT OF ANIMATION AND VISUALIZATION MODELS

The spatial and hydrological data sets are used to complete the task of animation. The data obtained for the region is evaluated for its use to develop the simulation of the catastrophic flood event. The main objective of this task is to identify and select a critical region of South Florida that is vulnerable to catastrophic flooding events. The selection will depend on the availability of spatial and temporal water level data associated with such events. The water level data is provided by the local water management agency, South Florida Water Management District, based on results from their in-house hydrologic simulation combined with an approximate method of flood stages in the selected region for a pre-selected extreme rainfall event. The spatial data related to the extent defined by the coordinates are then collected and compiled. Several data and geoprocessing tasks that are carried out are shown in Figure 4.2 and are also described in the following sections.

4.6 CREATION OF TERRAIN DATABASES/ FILES FOR 3D CULTURE AND IMAGERY

A mosaic of DOQQs (digital ortho quarter quadrangles) is created with a reduced resolution of each tile. The maximum resolution of each tile is fixed at 2048 × 2048 pixels. The mosaic is then converted into a GeoTIFF. The georeferencing of aerial images is carried using the Global Mapper software. The digital elevation model (DEM) and flooding depth layer were converted into a format appropriate for the animation software. The proprietary software used for this study provides three modeling tools that help to edit vector and raster data and terrain processing capabilities. The terrain creation process involves the development of a terrain grid using one of the tessellations: Polymesh and Delauney. In this study, Polymesh tessellation (DeMers, 2002) comprising of a uniform rectilinear grid of triangles was used to set up terrain for the animation. The Polymesh tessellation is shown in Figure 4.3. Details of the spatial data used for the study are provided in Table 4.1. The software used for the creation of terrain and 3D culture imagery is MultiGen Creator studio.

4.7 GENERATION OF TOPOGRAPHY FOR USE IN ANIMATED ENVIRONMENT

The topography of the region is depicted using a DEM and is converted into USGS (United States Geological Survey) DEM format before it is imported into the animation software. The DEM is finally converted into a specific format referred to as the DED (Digital Elevation Data) format acceptable for the software to generate the topography of the area. The Digital Elevation Data (DED) format is a uniform elevation map format that is used for terrain building by the software. The proprietary software used in the study includes several utilities for converting standard file formats, such as the United States Geological Survey (USGS) Digital Elevation Model (DEM) format as well as the United States National Imagery and Mapping Agency (NIMA) Digital Terrain Elevation Data (DTED) format, to DED format.

The infrastructure and features (roads, office buildings, houses, etc.) are developed using available information about their location and structural details. The buildings and other features are created as objects within the modeling environment and are overlaid on the aerial imagery layer. Textures (2D images defined as rectangular arrays of color data elements) are then mapped on to 3D shapes. An example of texture can be seen in in Figure 4.4.

Figure 4.3 Polymesh tessellation developed and used for the study.

Table 4.1 Spatial, Infrastructure, and Hydrological Data Sets Used for the Study

Data Type	Data Source
Spatial	
Digital Ortho Quarter Quadrangles (DOQQ)	Google Earth
Digital Elevation Model (DEM)	www.fgdl.orgRaster cell resolution: 30 × 30 m [USGS National Elevation Dataset (NED)] – Vertical Datum – NGVD 29
Infrastructure	
SFWMD Canal Structure Data	SFWMD
SFWMD Hydraulic Structures	SFWMD
Hydrological Data	
Water Level Layer (inundation depth) for a hypothetical event	SFWMD

SFWMD: South Florida Water Management District, Florida, USA.

FGDL: Florida Geographic Data Library.

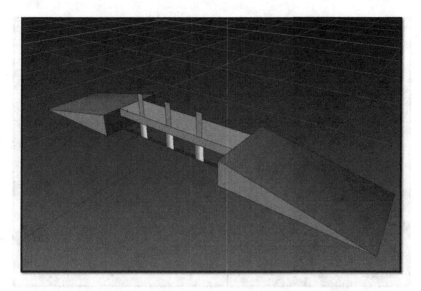

Figure 4.4 Three-dimensional (3D) drawing of a hydraulic structure.

Water canal structures are also created based on available details, as shown in Figure 4.5. Due to computational resource constraints associated with the 3D depiction of several hydraulic structures, only the locations are identified by point features in the animation sequence. Buildings and houses and all other features are created at two locations in the region to demonstrate the capabilities of the software in depicting the features in 3D. In many instances, photographs of the buildings can be used to develop these textures. However, the use of too many textures in the animation can place a burden on computational resources. Details of the infrastructure can be found in Teegavarapu et al. (2007) (Figure 4.6).

4.8 SPATIAL DATA RESOLUTION AND COMPATIBILITY ISSUES

Spatial data acquired from several different sources raises compatibility and computational issues. The proprietary software used in the current study for animation utilizes a specific format for spatial data, especially for the digital elevation model. The computational burden related to using high-resolution aerial imagery forced the study team to seek other sources such as "Google Earth" to acquire the data at low resolution. Considering the spatial extent of the region (i.e., South Florida) that is being modeled, the resolution of DEM and aerial imagery was restricted to manage the available computational resources for animation. The limitations of the proprietary

Figure 4.5 Series of hydraulic structures in the study region.

Figure 4.6 Aerial extent of flooding in South Florida caused by a hypothetical extreme hurricane event.

software in handling large amounts of spatial data need to be recognized. The hypothetical flooding depth (raster data layer) information provided by a local water management agency was initially used to model the extent of flooding in the selected region. The use of this layer in the animation software revealed some issues and problems in accurately depicting the depth of flooding. These issues were related to datum and resolution of the data layers. Finally, the inundation raster data was used to address the issue.

4.9 LEVEL OF DETAIL (LOD)

The computation resources available and the resolution of the spatial data sets define the level of detail (LOD) that can be incorporated into an animation scheme. In computer graphics literature, accounting for the level of detail involves decreasing the complexity of the 3D image as the viewer moves away from the object. By incorporating the level of detail techniques, the efficiency of the graphics display in the development of raster images can be improved. The reduced visual quality of the image at larger distances from the object of interest may not compromise the detail in animation at that level. Switching distances are defined in the animation scheme, where the level of detail switches from one to another. The level of detail automatically switches at a specific distance (from the center of the model to the eyepoint). The models are observed at varying distances from the eyepoint, and for a specific distance when the difference between two models at a certain distance cannot be discerned, that distance is specified as switching distance. The number of polygons used to display the detail is reduced by approximately 60% when animation switches from a higher level of detail to a lower level of detail. In the current study, the level of detail concept is used to define the details of the topography.

4.10 Z-BUFFERING EFFECT

Z-buffering[1] is an algorithm used in 3D graphics to ensure that perspective works the same way in the virtual world as it does in the real one. Z-buffering works by testing pixel depth and comparing the current position (z coordinate, in the current context, the flooding depth) with stored data in a buffer (called a *Z-buffer*) that holds information about each pixel's last position. It automatically computes which pixels are at higher depth and which are at lower depth and helps the computing environment to identify which surfaces need not be drawn since they are below other surfaces. Depending on the type of modeling environment adopted, video graphics card used, and the available computational resources, the Z-buffering process may not successfully differentiate the layers of information. This is referred to as the Z-buffer problem.

In this pilot study, a Z-buffer problem was encountered when large data sets (related to topography and flooding depth) were overlaid on top of each other and visualized in the animation mode. However, when animation closer to the object of interest is performed, the Z-buffer problem surfaces. This is shown by the inaccurate characterization of the topography and the flooding depth. To circumvent the Z-buffer problem and to clearly distinguish between the topography layer and the flooding depth layer, objects (vertical staffs) are created and placed at a few select regions of South Florida. These objects will provide the user with the animation to identify and assess the flooding depths.

4.11 FLY-BY ANIMATION

Animation software is used to develop 3D models of the data. The creation of multiple scenarios/simulations for a different combination of meteorological parameters (e.g., wind speed, daylight conditions, and others) is possible by using the appropriate software. A "fly-by" refers to the creation of an animation sequence based on the directed flight path on top of the area of interest. The creation of AVI (Audio Video Interleave) files for a specific scenario is possible. The extent of aerial flooding and depth can be shown by a fly-by animation scheme.

4.12 LIMITATIONS

Several inherent limitations of the software and datasets that are used for the development of 3D animation of the catastrophic flooding event should be noted. The main limitations are due to the computational burden of simulating processes at a large areal scale. High-resolution (fine resolution) data sets (DOQQs, DEM, and others) require large computational memory for the execution of the animation. The animation software used for this study is more geared toward simulation and animation of small regions with extensive detail. Several recommendations are made based on this study, and they include: (i) reduction of the spatial resolution of the data sets used for the study will help in making the simulation and animation manageable with a maximum level of detail and also based on available computational resources; (ii) phased simulation/animation of different regions is recommended as opposed to modeling one complete region to reduce the computational resource requirements and to avoid execution time constraints; and (iii) high-resolution spatial data sets are required at several crucial locations to study the flooding effects in more detail. The study reported in this chapter deals with the simulation and animation of catastrophic flood event conditions. The methodology provides a new interactive animation mechanism to study extreme flooding events. While the case study region to which the methodology applied is specific, the approach is generic and can be transferable to any region. The study provides an interactive environment for flood damage assessment with the exception of a few minor issues raised by types of data sets used, computational resources, and execution time-related problems. The level of detail presented in the animation scheme will enable the decision-makers at the water management agency to understand and assess the extreme flooding conditions in space. This will also help water managers and emergency personnel to develop evacuation plans. Several improvements to the animation can be achieved by setting a clear focus on smaller regions using spatial data of finer resolution.

4.13 SUMMARY AND CONCLUSIONS

The development of a simulation model aided by the animation approach is briefly discussed in this chapter. Simulation with animation can aid in the evaluation of different disaster management scenarios. An appropriate extension of the approach presented in this chapter would be to enhance the simulation using virtual reality tools for users to explore management options under immersive modeling environments. The next two chapters will discuss virtual reality modeling approaches that can be used to model water resources systems and help in disaster management.

Note

1. Source: Whatis.com, TechTarget.

GLOSSARY

Animation	It is the rapid display of a sequence of images of 2D artwork or model positions to create an illusion of movement
GeoTIFF	It is a public domain metadata standard which allows georeferencing information to be embedded within a TIFF file
Shapefile	It is a popular geospatial vector data file format for geographic information systems software developed by Environmental Science and Research Institute (ESRI)
Z-buffer	In a computer graphics card, this section of video memory keeps track of which onscreen elements can be viewed and which are hidden behind other objects (source: https://docs.unrealengine.com/udk/Three/UnGlossary.html)

REFERENCES

Ashley, R., S. Garvin, E. Pasche, A. Vassilopoulos, C. Zevenbergen (editors) (2007). *Advances in Urban Flood Management*. Taylor and Francis, The Netherlands.

Bates, P. D., A. P. J. De Roo (2000). A simple raster-based model for floodplain inundation. *Journal of Hydrology*, 236, 1–2, 54–77.

Bell, J. T., H. S. Fogler (1998). *Virtual Reality in Chemical Engineering Education, ASCEE*. Northcentral section meeting, University of Detroit Mercy, Detroit, MI.

Dalponte, D., P. Rinaldi, G. Cazenave, E. Usunoff, L. Vives, M. Varni, M. Vénere, A. Clausse (2006). A validated fast algorithm for simulation of flooding events in plains. *Hydrological Processes*, 21(8), 1115–1124.

De Mers, M. N. (2002). *GIS Modeling in Raster*. John Wiley. New York.

Garcia, M. J. (2018). *Theory and Practical Exercises of System Dynamics*. Independently published.

Gyorfi, J. S., E. R. Buhrke, M. A. Tarlton, J. M. Lopez, G. T. Valliath (2006). *VICC: Virtual Incident Command Center*. Proceedings of 9th Annual International Workshop on Presence, Presence 2006, International Society of Presence Research, Philadelphia.

Horritt, M. S., P. D. Bates (2001). Predicting floodplain inundation: Raster-based modelling versus the finite element approach. *Hydrological Processes*, 15(5), 825–842.

MacFarlane, R. (2005). *A Guide to GIS Applications in Integrated Emergency Management*. Emergency Planning College, Cabinet Office.

Maidment, D. R., O. Robayo, V. Merwade (2005). *Hydrologic Modeling, GIS, Spatial Analysis and Modeling*. ESRI Press, Redlands, California.

Pajorova, E., E. Hluchy, J. Astalos (2007). *3D Geovisualization Service for Grid-Oriented Applications of Natural Disasters*. Krakow 07 Grid Workshop, Krakow, Poland.

Teegavarapu, R. S. V., P. Scarlatos, Y. Kaner (2007). *A Pilot Study on Catastrophic Flood Scenario Animation for a Region in South Florida*. SFWMD Technical Report, Center for Inter-modal Transportation Safety and Security, South Florida Water Management District (SFWMD), West Palm Beach, Florida, 38p.

Romanowicz, R., K. Beven (2003). Estimation of flood inundation probabilities as conditioned on event inundation maps. *Water Resources Research*, 39(3), 1061–1073.

Sampio, A. Z., C. O. Cruz, O.P Martins (2011). Didactic models in civil engineering education: Virtual simulation of construction works. *Virtual Simulation of Construction Works*, Jae-Jin Kim, ISBN: 978-953-307-518-1, InTech.

Sene, K. (2008). *Flood Warning, Forecasting and Emergency Response*. Springer, The Netherlands.

Sene, K. (2012). *Flash Floods: Forecasting and Warning*. Springer, The Netherlands.

Van Oosterom, P., S. Zlatanova, E. M. Fendel (editors) (2005). *Geo-Information for Disaster Management*. Springer, Berlin.

Chapter 5

Basics of Virtual Reality Modeling

John Moreland, Chenn Q. Zhou, and
Chandramouli V. Chandramouli

5.1 INTRODUCTION

Virtual reality (VR) is a combination of technologies used to create computer-generated three-dimensional, immersive environments that let users
experience an artificial version of reality. VR typically includes a combination of stereoscopic 3D display, viewer-centered perspective, interactive and
real-time experiences, and various multisensory inputs and outputs such as
hand tracking. User experiences in VR typically mimic real-world experiences but remove limitations that are based on physical world constraints,
enabling users to fly around within a 3D environment and interact with
complex data.

The resulting virtual environments can be used for a variety of applications ranging from entertainment to education and can be blended with
three-dimensional simulation modeling results to provide useful inferences
and a clear understanding of the processed results (Fu et al., 2009, p. 1;
Sampaio, Henriques, and Martins et al., 2010, p. 18; Viswanathan et al.,
2011, p. 1). In engineering, for complex problem-solving, simulation models are developed in a 2D, quasi 3D, or 3D modeling environment. Vast
expertise and skill are often necessary to interpret and implement such
models. VR tools substantially simplify these problems and create user-
friendly animations and visualization (Libin, 2001, p. 652). Using VR in
these cases helps new users and provides scope for anyone to use the results
more effectively.

5.1.1 Emergence – Historic Perspective

The term 'virtual reality' was originally coined by Jaron Lanier, founder of
VPL (Virtual Programming Language) Research, in the 1980s (Hamilton
et al., 1992, p. 652). However, the various forms of VR stretch back to
the 1950s with the Sensorama and Ivan Sutherland's "Ultimate Display"
(Figure 5.1), and some elements of VR could be seen as early as the 1860s
with the introduction of stereoscopic photographs (Figure 5.2) (Heilig,
1962; Sutherland, 1965; NH 59429-A).

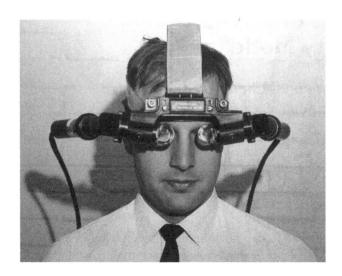

Figure 5.1 Ivan Sutherland's "Ultimate Display."

Photo # NH 59429-A   RAdm. John A. Dahlgren and his staff, on board USS Pawnee -- Stereo pair

Figure 5.2 Early stereoscopic 3D photograph, circa 1860s.

While these early efforts focused on VR for individual users using head-mounted displays (HMDs), the introduction of the CAVE (CAVE Automatic Virtual Environment) in the early 1990s introduced multi-user projection-based environments (Cruz-Niera et al., 1992, p. 64). Projection-based VR then dominated much of VR research for several decades (Figure 5.3). In the early 2010s, early versions of the Oculus Rift drastically improved the affordability of high-quality HMDs (Figure 5.4), beginning the modern era of VR, which has brought the technology into the mainstream (Rubin, 2014; Bown et al., 2017, p. 239).

Figure 5.3 Projection-based VR system.

Figure 5.4 Oculus Rift HMD (head-mounted display) VR system.

The success of VR in the entertainment industry has been instrumental in improving the technologies to better apply them in engineering applications and problem-solving (Petukhov et al., 2017, p. 1). VR applications are well suited for situations where real-time training is difficult or dangerous or inhospitable ground conditions. So the scientific community started using VR for those applications such as space training and pilot training. Recent students are more used to several electronic gadgets and technological tools from a young age. Using cutting edge training in college education is mandatory to prepare them well for industry (Abulrub et al., 2011, p. 751). In recent times, computer hardware and software capabilities help us to use highly interactive virtual 3D models in engineering education (Messner et al., 2003, p. 1). VR models make human interaction to the process by providing friendly connectivity to the process involved (Mujber et al., 2004, p. 1834). With the advent of computing capabilities and memory available, VR applications were explored by many researchers in different areas of research such as engineering, technology, medical applications, safety training, and education (Kerawalla et al., 2006, p. 163; Smeets, 2005, p. 343).

5.1.2 Computing Requirements

VR applications and implementations typically have high computing and storage requirements. Bryson (1996, p. 63) presented the complexities involved in three-dimensional time series data. Rendering ability suffers when sufficient memory is not available (Schaufler and Stürzlinger, 1996, p. 227; Oliveira et al., 2007, p. 1715). In an interactive VR model, the speed of the model rendering process is also critical (Oliveira et al., 2007, p. 1715). Images and the virtual space and 3D models generated by the operating platforms require significant disk storage and virtual memory space. However, novel approaches in application development also reduce memory needs (Hanvu and Yueming, 2009, p. 686). Seth et al. (2011, p. 5) provided a comprehensive review of VR prototyping. They indicated the uses of the virtual assembly process and the difficulties to develop complex models. Even though continuous advances and improvements in computing, helps the users, the demands of users are also keep increasing to have a more real world feel with the virtual reality. It increases the size and complexity of VR simulations. To run the developed application with high resolution, sufficient random access memory (RAM) is essential or one should use parallel architecture (Bryson, 1996, p. 63).

5.1.3 Benefits of VR

There are inherent benefits to the use of virtual reality. With multiple tools available in the market, VR can be used in generating simulated scenes in near real time. By properly designing the expected output, VR can be used to generate many processes (Manseur, 2005, p. F2). Manseur (2005, p. F2) also states, "It is now possible to illustrate complex, expensive, or danger-ous systems safely and economically on a computer screen." For providing a clear understanding of a futuristic scenario, VR is very useful. For example, by adopting proper design, a multi-million dollar dam construction and reservoir at a considered site and the expected inundations can be presented to a future investor easily using VR. It can also be developed with different types of spillways, and the client can view different perspectives easily and understand the implications.

VR applications are popularly used in education and training in many fields, including medical education (Preece et al., 2013, p. 216). When it becomes practically impossible for a human to get to the site, virtual reality becomes a handy option. For example, if we have a blast furnace operat-ing at extreme temperatures, a change in the process leading to an inferior product formation cannot be inspected by a human directly, but integrating simulation data, sensors, and virtual reality can allow operators to under-stand and make appropriate decisions (Wang et al., 2014, p. 1; Fu et al., 2009, p. 307). In a virtual environment, we can simulate that situation and use it as a training tool in the workplace. Another important advantage of this technique is a versatile nature. With few or large modifications, varia-tions in scenarios can be explained or presented very well to the operators to train them in severe conditions and the system failures (Manseur 2004, p. 1). With advancements in information technology, web-based applica-tions make it possible to share VR models through the web. The resources created can be used for remote learning with high-quality computer graph-ics contents (Monahan et al., 2008, p. 1339). All these advantages have led to a wide range of applications of VR in different fields (Figure 5.5).

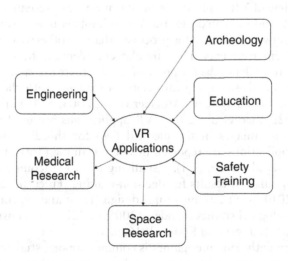

Figure 5.5 VR applications in various fields.

5.1.4 Limitations of Visualization of 2D or Quasi 3D Model Simulations

Even though there were many applications of VR in computer games, image visualization, and other fields which don't involve geometric data, this work is mainly focused on presenting the three-dimensional modeling results in real-world applications in different engineering fields (Mansuer, 2004, p. 1). In most of the engineering applications, when real-world system modeling is done in a three-dimensional environment, temporal variations are also included as a fourth dimension in the modeling domain (Chandramouli et al., 2016, p. 934). An enormous amount of data needs to be captured in an orderly fashion and to be arranged systematically to create VR applications. However, many real-world applications are still handled in a 1D, 2D, or quasi-3D framework. These decisions were taken usually based on previous studies and their successes in interpreting them in real-time systems (Vojinovic and Tutulic, 2009, p. 183). Popular software available in the public domain, data discrepancies, cost, and computational requirements are other factors. But certainly, they will bring in compromises based on the problem considered because it leads to interpolating data to the next dimension or making approximations based on boundary conditions (Vozinaki et al., 2017, p. 642). The advantages of using a 1D model, as well as the benefits derived from 2D modeling while handling a complex flood plain, were presented by Vozinaki et al. (2017, p. 642).

5.1.5 Applications as a Teaching Aid

Even though VR technology is applied in different fields, it is very popularly used as a training and teaching tool (Bowmen et al., 1999, p. 317). A detailed review of VR applications in education was presented by Manseur (2005, p. F2) and indicated that the VR technology is very suitable to create an interactive laboratory for processes that are otherwise dangerous or inhospitable environments. They are also considered to be a good alternative to regular student labs for processes when there are space restrictions. VR facilitates active learning and engages students with the appropriate design of considered systems (Messner et al., 2003, p. 1; Bailenson et al., 2008, p. 102; Zhao et al., 2017). VR applications are used in analyzing and visualizing images in the medical field for the diagnostic process. Extensive applications are reported in safety training (Toth et al., 2019, p. 1; Moreland et al., 2019, p. 1), visualizing the CFD (computational fluid dynamics) simulation results for decision-making (Fu et al., 2010, p. 761; Wu et al., 2010, p. 51), architectural design (Frost and Warrant, 2000, p. 568), archaeological studies (Sanders, 2008, p. 427), and construction and space research (Hilfert and Konig, 2016, p. 1).

Familiarity with computer games is common among students and translates to the easy use of interactive applications to enhance learning. VR technology with 3D results can be used together to generate complex shapes to help passive learners (Sampaio et al., 2011).

5.2 ESSENTIAL VR DEVELOPMENT TOOLS

While dedicated VR development tools have emerged through the years to meet the needs of VR researchers such as Vizard and VR Juggler (WorldViz, 2012; Bierbaum et al., 2001, p. 89), the overwhelming majority of modern VR applications are now developed using tools that were created for the gaming industry. Visual 3D models are typically developed using programs such as 3ds Max, Maya, and Blender and then brought into game engines such as Unity 3D and Unreal Engine to combine with other assets to develop the final interactive VR application (Jerald et al., 2014, p. 1; McCaffrey, 2017).

5.2.1 Introduction: Different Tools Available

5.2.1.1 Unity 3D

Unity 3D (https://unity.com/) is currently the most popular game engine for VR developers and provides a framework and tools to develop interactive software that can be published to multiple platforms including personal computer (PC), Mac, mobile devices, web, VR, and AR (augmented reality). A user-friendly interface allows basic interactive applications to be built with minimal programming experience; more complex VR applications may still require coding to produce desired interactive experiences. C# is the coding language used to develop scripts in Unity. An online asset store is integrated with the software and acts as a community-driven marketplace for Unity developers to buy or sell assets to use in their projects such as 3D models, animations, programming scripts, and even full game templates for users to customize and publish games and VR applications with minimal effort. Unity offers both a free Personal and paid Professional version of the software, with the main difference being whether or not the user can sell their application.

5.2.1.2 Unreal Engine

The Unreal Engine (www.unrealengine.com/) provides most of the same or similar functionality for VR developers as Unity 3D but with several notable differences. One key difference is the use of a "blueprint" visual-scripting system, which allows developers to create games and VR applications with zero coding. The blueprint system uses a visual interface to connect preconstructed pieces that take the place of scripting code. There are various other differences, and both software packages have hardcore users and fan bases claiming superiority, but you can largely create comparable applications using either software. Due to licensing/cost structure, Unreal is typically more associated with big-budget game development, while Unity is more associated with mobile and small-budget games.

5.2.1.3 3D Modeling

Another key piece of VR development is the creation of 3D models and animations. While both Unity 3D and Unreal provide some tools to create simple models, and it is often possible to find 3D models available for free or for purchase online; a dedicated 3D modeling software is sometimes needed to create custom 3D models. While there are many options for modeling software, 3ds Max, Maya, and Blender are among the more popular 3D modeling packages (Gahan, 2013; Labschutz et al., 2011, p. 124; Hess, 2007). Anything you can create in one software you can also create it in one of the others. They each provide similar functionality and capability but with slightly different interfaces. 3ds Max is more associated with architectural modeling and artificial structures, while Maya is often favored for creating more organic shapes and character models. Both 3ds Max and Maya are commercial software, while Blender is an open-source "free" option. Regardless of which software is used, 3D models are typically created, textured (by wrapping photos and images around the 3D model geometry), and then exported to a common 3D file format such as FBX. The exported 3D models can then be imported into the game engine for use in the VR software.

Kim et al. (2014, p. 21) discussed the details of developing augmented reality applications using the Unity 3D platform. The essential aspect of VR application is user interactive features. Tools such as 3ds Max and Unity 3D platform offer flexible built-in modules, objects, and 3D tools (Indraprastha and Shinozaki, 2009, p. 1; Miyata et al., 2010, p. 811). They also provide standard 3D navigation tools to support VR developments. These tools also minimize coding and provide flexibility to the application developers to explore and improve the quality of VR with interactive features and better rendering effects. The selection of a particular tool is based on factors such as the ability to build intricate 3D models and create virtual scenes, rendering capabilities, user-interactivity creation tools, flexibility to add textures, optimal size files, and reasonable memory requirements.

5.3 DEVELOPMENT OF VR APPLICATIONS

Development of VR applications for a considered process begins with a clear understanding of the process and identifying objectives that the VR application will meet. Once the objectives are decided, factors such as the targeted scene of the system, how to navigate through the system, how users should

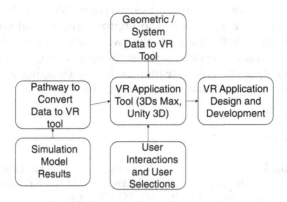

Figure 5.6 VR application development architecture.

interact with objects in the scene, and the expected interaction-based changes in the application should be brainstormed (Chandramouli et al., 2016). 3D models, data, and other items which will be included in the scene should then be developed and imported into the development environment to blend for the VR application (Ozvoldova et al., 2006, p. 297) (Figure 5.6).

In some processes which include simulation or sensor data, temporal and spatial changes may also need to be planned for and implemented in the application to allow users to see a process change over time. In these cases, controls should be developed to enable the user to change time using common mechanisms such as a slider, which can be dragged around to select a specific time step, or automated to show changes occur smoothly over a designated period. This type of temporal navigation can result in dynamic changes in the generated environment in VR. Because the 3D simulation results are usually vast, the proper pathway to convert the data in the required format is a very crucial step. Those pathways will be very specific for each case. Other effects such as navigation in the virtual environment, direction identification during navigation, camera positioning, and measurement tools while using the application are also needed and should be included in planning (Shen and Zeng, 2011, p. 487). Once the plan is in place, the application should be developed and tested by both the development team and users who are unfamiliar with the software. This will help to identify any usability issues or points of confusion. The software should be revised until a satisfactory state is achieved, at which point it can be released for users to access.

5.4 SUMMARY AND CONCLUSIONS

Virtual reality has developed into a very robust and useful technology with numerous useful applications. In recent years, advances in computing and the availability of "cheap" VR hardware have enabled the development of many educational VR applications to improve student understanding and learning outcomes. Software tools that were originally developed by the gaming industry now provide a user-friendly option for developing VR software. This enables the rapid creation of highly engaging and immersive learning experiences for students in multiple disciplines.

REFERENCES

Abulrub, A. G., A. N. Attridge, M. A. Williams (2011). Virtual reality in engineering education: The future of creative learning. *2011 IEEE Global Engineering Education Conference (EDUCON)*, 751–757, Amman.

Bailenson, J. N., N. Yee, J. Blascovich, A. C. Beall, N. Lundblad, M. Jin (2008). The use of immersive virtual reality in the learning sciences: Digital transformations of teachers, students, and social context. *Journal of the Learning Sciences*, 17(1), 102–141.

Bierbaum, A., C. Just, P. Hartling, K. Meinert, A. Baker, C. Cruz-Neira (2001). VR Juggler: A virtual platform for virtual reality application development. *Proceedings IEEE Virtual Reality 2001*, 89–96, Yokohama, Japan. doi:10.1109/VR.2001.913774.

Bowman, D. A., L. F. Hodges, D. Allison, J. Wineman (1999). The educational value of an information-rich virtual environment, Presence. Teleoperators and Virtual Environments, 8(3), 317–331.

Bown, J., E. White, A. Boopalan (2017). Looking for the ultimate display: A brief history of virtual reality. In J. Gackenbach, J. Bown, *Boundaries of Self and Reality Online: Implications of Digitally Constructed Realities*, 239–259). Elsevier Academic Press. https://doi.org/10.1016/B978-0-12-804157-4.00012-8

Bryson, S. (1996). Virtual reality in scientific visualization. *Communications of ACM*, 39(5), 63–71.

Chandramouli, C. V., Z. Ziong, M. J. Wang, J. Moreland, C. Zhou, R. S. V. Teegavarappu, J. Fox, P. Behera, E. Hixon (2016). Virtual 3D modeling for little calumet river flood inundation studies. EWRI Congress 2016, May 2016 at West Palm Beach, FL.

Cruz-Niera, C., D. J. Sandin, T. A. DeFanti, R. V. Kenyon, J. C. Hart (1992). The CAVE: Audio visual experience automatic virtual environment. *Communications of the ACM*, 35(6), 64–73.

Frost, P., P. Warren (2000, July). Virtual reality used in a collaborative architectural design process. In 2000 IEEE Conference on Information Visualization. International Conference on Computer Visualization and Graphics, 568–573, IEEE.

Fu, D., B. Wu, G. Chen, J. Moreland, F. Tian, Y. Hu, C. Zhou (2010). Virtual reality visualization of CFD simulation for iron/steelmaking processes. *2010* 14th *International* Heat Transfer Conference *(IHTC14)*, 761–768, Washington, DC.

Fu, D., B. Wu, J. Moreland, G. Chen, S. Ren, C. Zhou (2009). Virtual reality visualization of typical processes in blast furnace. *2009 AISTech - Iron and Steel Technology Conference Proceedings*, Vol. 1, 307–314, St. Louis, MO.

Fu, D., B. Wu, J. Moreland, G. Chen, C. Zhou (2009). CFD simulations and VR visualization for process design and optimization. *Proceedings of the Inaugural US-EU-China Thermophysics Conference*, Beijing, China, UECTC-RE '09, UECTC-RE T5-S6-029.

Gahan, A. (2013). *3ds Max Modeling for Games: Insider's Guide to Game Character, Vehicle, and Environment Modeling.* CRC Press, Oxford, UK.

Hamilton, J., E. T. Smith, G. McWilliams, E. I. Schwartz, J. Carey (1992). Virtual reality: How a computer-generated world could change the real world. *Business Week*, 3286, 97–104.

Hanwu, H., W. Yueming (2009). Web-based virtual operating of CNC milling machine tools. *Computers in Industry*, 60(9), 686–697. https://doi.org/10.1016/j.compind.2009.05.009

Heilig, M. L. (1962). U. S. Patent No. 3,050,870. U.S. Patent and Trademark Office, Washington, DC.

Hess, R. (2007). *The Essential Blender: Guide to 3D Creation with the Open Source Suite Blender.* No Starch Press, San Francisco.

Hilfert, T., M. König (2016). Low-cost virtual reality environment for engineering and construction. *Visualization in Engineering*, 4(2), 1–18. https://doi.org/10.1186/s40327-015-0031-5

Indraprastha, A., M. Shinozaki (2009). The investigation on using unity 3D game engine in Urban design study. *Journal of ICT Research and Applications*, 3(1), 1–18.

Kerawalla, L., R. Luckin, S. Seljeflot, A. Woolard (2006). Making it real: Exploring the potential of augmented reality for teaching primary school science. *Virtual Reality*, 10(3–4), 163–174. Springer-Verlag London Ltd, London.

Kim, S. L., H. J. Suk, J. H. Kang, J. M. Jung, T. H. Laine, J. Westlin, (2014). Using unity 3D to facilitate mobile augmented reality game development. *2014* IEEE World Forum *on* Internet *of* Things *(WF-IoT)*, 21–26, Seoul. doi:10.1109/WF-IoT.2014.6803110.

Labschütz, M., K. Krösl, M. Aquino, F. Grashäftl, S. Kohl (2011). Content creation for a 3D game with Maya and Unity 3D. *Institute of Computer Graphics and Algorithms*, Vienna University of Technology, 6, 124.

Libin, A. (2001, October). Virtual reality as a complex interactive system: A multidimensional model of person artificial partner co-relations. *Proceedings Seventh International Conference on Virtual Systems and Multimedia*, 652–657. IEEE.

Manseur, R. (2004). Interactive visualization tools for robotics. *Proceedings of the Florida Recent Advances in Robotics*, 1–7, Orlando, FL.

Manseur, R. (2005). Virtual reality in science and engineering education. *Proceedings Frontiers in Education 35th Annual Conference*, F2E-8, Indianapolis, IN.

McCaffrey, M. (2017). *Unreal Engine VR Cookbook: Developing Virtual Reality with UE4.* Addison-Wesley Professional, Boston, USA.

Messner, J. I., S. C. M. Yerrapathruni, A. J. Baratta (2003). Using virtual reality to improve construction engineering education. *Proceedings of the 2003 American Society for Engineering Education Annual Conference & Exposition.*

Mikropoulos, T. A., A. Natsis (2010). Educational virtual environments: A ten-year review of empirical research (1999–2009). *Computers & Education*, Elsevier, 56, 769–780.

Miyata, K., K. Umemoto, T. Higuchi, (2010). An educational framework for creating VR application through groupwork. *Computers & Graphics*, 34(6), 811–819.

Monahan, T., G. McArdle, M. Bertolotto (2008). Virtual reality for collaborative e-learning. *Computers and Education*, Elsevier, 50(4), 1339–1353.

Moreland, J., K. Toth, Y. Fang, M. Block, G. Page, S. Crites, C. Zhou (2019). Interactive simulators for steel industry safety training. *Steel Research International*, 90, 1800513, 1–9. doi:10.1002/srin.201800513

Mujber, T. S., T. Szecsi, M. S. Hashmi (2004). Virtual reality applications in manufacturing process simulation. *Journal of Materials Processing Technology*, 155, 1834–1838.

NH 59429-A courtesy of the Naval History & Heritage Command.

Oliveira, D. M., S. C. Cao, X. F. Hermida, F. M. Rodriguez (2007). Virtual reality system for industrial training. *IEEE Explore*, 1715–1720. doi:10.1109/ISIE.2007.4374863

Ozvoldova, M., P. Cernansky, F. Schuer, F. Lustig (2006). Internet remote physics experiments in a student laboratory. *Innovations 2006*, Vol. 25, 297–304, W. Aung, et al., iNEER, Arlington, VA.

Petukhov, I., L. Steshina, A. Glazyrin (2017, November). Application of virtual reality technologies in training of man-machine system operators. *2017 International Conference on Information Science and Communications Technologies (ICISCT)*, 1–7. IEEE.

Preece, D., S. B. Williams, R. Lam, R. Weller (2013). *"Let's Get Physical": Advantages of a Physical Model over 3D Computer Models and Textbooks in Learning Imaging Anatomy*. Research Report, Anatomical Science Education, American Association of Anatomy, 6, 216–224. https://doi.org/10.1002/ase.1345

Rubin, P. (2014). *The Inside Story of Oculus Rift and How Virtual Reality Became Reality* [online] Wired. com.

Sampaio, A. Z., C. O. Cruz, O. P. Martins (2011). Didactic models in civil engineering education: Virtual simulation of construction works. *Virtual Simulation of Construction Works*, Prof. Jae-Jin Kim, InTech. ISBN: 978-953-307-518-1

Sampaio, A. Z, P. G. Henriques, O. P. Martins (2010). Virtual reality technology used in civil engineering education. *The Open Virtual Reality Journal*, 2010(2), 18–25.

Sanders, D. (2008). Why do virtual heritage? *Digital Discovery: Exploring New Frontiers in Human Heritage Proceedings from the 34th Computer Applications and Quantitative Methods Archaeology Conference*, 427–436, J. T. Clark, E. M. Hagemeister, Fargo, ND, April 2006, Budapest: Archaeolingua.

Schaufler, G., W. Stürzlinger (1996). A three dimensional image cache for virtual reality. *Computer Graphics Forum*, 15(3), 227–235. doi:10.1111/1467-8659.1530227

Seth, A., J. M. Vance, J. H. Oliver (2011). Virtual reality for assembly methods prototyping: A review. *Virtual Reality*, 15, 5–20. doi:10.1007/s10055-009-0153-y

Shen, W., W. Zeng (2011). Research of VR modeling technology based on VRML and 3DSMAX. *Proceedings of 2011 International Conference on Computer Science and Network Technology*, 487–490, Harbin. doi:10.1109/ICCSNT.2011.6182002

Smeets, E. (2005). Does ICT contribute to powerful learning environments in primary education? *Computers & Education*, 44(3), 343–355.

Sutherland, I. E. (1965). The ultimate display. *Proc. IFIP*, 65(2), 506–508, 582–583.

Toth, K., J. Moreland, F. L. Zhang, A. Balachandran, J. Roudebush, S. Vietor, C. Zhou (2019). Development of an educational wind turbine troubleshooting and safety simulator. *2019 American Society for Engineering Education (ASEE) Annual Conference & Exposition*, Columbus, OH.

Viswanathan, C., J. Moreland, S. Guo, C. Zhou (2011). Usefulness of virtual 3D modeling to visualize the effect of uncertain data in groundwater solute transport. *Proceedings of the ASME 2011 World Conference on Innovative Virtual Reality*, WI.

Vojinovic, Z., D. Tutulic (2009). On the use of 1D and coupled 1D-2D modelling approaches for assessment of flood damage in urban areas. *Urban Water Journal*, 6(3), 183–199. doi:10.1080/15730620802566877

Vozinaki, A. K., G. G. Morianou, D. D. Alexakis, I. K. Tsanis (2017). Comparing 1D and combined 1D/2D hydraulic simulations using high-resolution topographic data: A case study of the Koiliaris basin, Greece. *Hydrological Sciences Journal*, 62(4), 642–656. doi:10.1080/02626667.2016.1255746

Wang, T., L. Phillips, J. Wang, D. Fu, J. Moreland, C. Zhou, Y. Zhao, J. Capo (2014). Development of a virtual blast furnace training system. *Materials Science & Technology 2014*, October 12–16, Pittsburgh, PA.

WorldViz, L. L. C. (2012). *Vizard Virtual Reality Toolkit*.

Wu, B., G. Chen, J. Moreland, D. Huang, D. Zheng, C. Zhou (2010). Industrial application of CFD simulation and VR visualization. *Proceedings of ASME World Conference on Innovative Virtual Reality*, 51–59, Ames, Iowa, WINVR2010-3734.

Zhao, J., P. LaFemina, J. O. Wallgrn, D. Oprean, A. Klippel, (2017). IVR for the geosciences. (2017). IEEE Virtual Reality Workshop *on* K-12 Embodied Learning *Through* Virtual Augmented Reality (*KELVAR*), 1–6, March.

Chapter 6

Virtual Reality Applications

Chandramouli V. Chandramouli, Chenn Q. Zhou, and John Moreland

6.1 INTRODUCTION

Virtual reality (VR) applications have shown success in multiple applications involving education and training. This chapter discusses multiple efforts that have been made to explore VR in flood modeling and related education (Chandramouli et al., 2016a, p. 1). VR offers many advantages such as versatility, rendering capabilities, and availability of advanced development tools. These were the key motivations for the development of a lab module to improve student understanding of flooding issues. Scenarios such as flooding cannot be developed in a laboratory environment due to issues such as space requirements, energy requirements, safety concerns, and cost aspects. However, using a virtual environment to carry out flooding labs is a convenient method to avoid such issues (Vergara et al., 2017, p. 1). The authors developed a VR flooding application that combined a virtual environment with hydrology and hydraulic modeling to provide a lab module that is both engaging for students and relevant to real-world flood scenarios.

Recently, due to global climatic changes, high-intensity rainfall events have been observed more frequently (Poff, 2002, p. 1488; Min et al., 2011, p. 378; Martino et al., 2013, p. 1). Severe weather conditions lead to large-scale flooding and bring great loss of property and changes in the extent of flood zones (Kirshen et al., 2008, p. 437). Flood modeling and flood plain analysis are important components in first- and second-level water resources engineering courses and are essential components of the civil engineering undergraduate curriculum. By integrating VR and flood plain modeling, a user-friendly application was developed to help students to understand the complex processes involved in flooding scenarios (Chandramouli et al., 2016b, p. 1). Advances in computing power and interactive 3D simulator capabilities have made real-time 3D flooding visualizations possible. Once created, these models can be easily modified for serving several tasks. Virtual reality has several exclusive characteristics such as being immersive, interactive, visually oriented, and exciting and is a particularly excellent tool for presenting three-dimensional objects and relationships (Bell and

Fogler, 1995, p. 1; Fu et al., 2009, p. 1). Details of the VR model development are presented in the next section.

6.2 VR FOR FLOODPLAIN INUNDATION MODELING

The objective of this modeling is to create an application and use it as an interactive lab module in a hydrology and hydraulics lab. An outline of the developed application is presented in Figure 6.1. A plan was also made to use it to evaluate the benefits of the virtual learning environment (Chandramouli et al., 2016b, p. 1). This model development process involves three distinct stages. The initial stage of this work involves hydrologic modeling, followed by the second stage, hydraulic modeling. The third stage of the model development is the interactive 3D virtual model, which uses the hydraulic and hydrologic model results. Each stage is explained in the following sub-sections.

6.2.1 Hart Ditch River System

The Hart Ditch River system is located in Northwest Indiana near Chicago, Illinois, in the U.S. (Figure 6.2). It is a man-made ditch constructed in the 1850s to drain many small towns located near the southern tip of Lake Michigan as a part of the land reclamation process. The system is located in the Lake Michigan watershed area. Hart Ditch flows from south to north and confluences with the Little Calumet River system, which flows east to west. This confluence point acts as a summit, and the water from Hart Ditch flows both eastward and

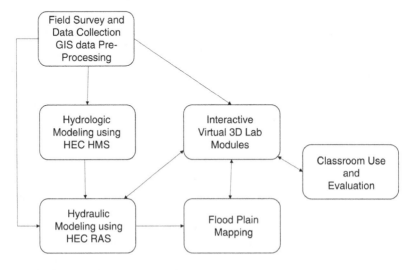

Figure 6.1 Outline of VR application developed.

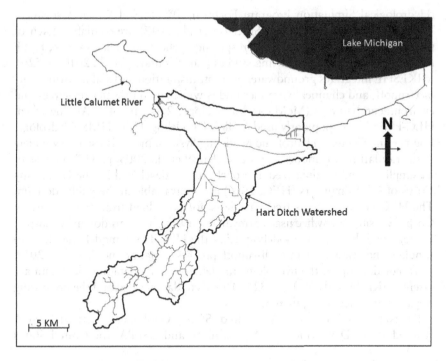

Figure 6.2 Little Calumet River watershed.

westward in the Little Calumet River. The western branch drains into Lake Michigan through Cull-de Sag Canal, and the eastern side drains into Lake Michigan through Burns Ditch. The Little Calumet River system is maintained by the Little Calumet River Commission. In 2008, a major flood was caused in this river basin resulting from Hurricane Ike. Huge property damages were reported due to this flood. The busiest interstate in the region, I-80-94, was closed for a week during this flood. In the aftermath of the flood, the U.S. Army Corps constructed levees and floodwalls along the Little Calumet River for a 22-mi stretch to avoid future flooding in Northwest Indiana (U.S. Army Corps Chicago District website). This system has a series of storage facilities created on the eastern side of the Little Calumet River. Both hydrology and hydraulics models were developed and calibrated for this system.

6.2.2 Hydrology Model Development

The flood modeling segment involves a two-stage modeling process, namely, hydrologic modeling (for the rainfall-runoff process) and hydraulic modeling for water movement through the channel (Thakur et al., 2017, p. 240) (Figure 6.1). A detailed review of available rainfall-runoff models was presented by Franchini and Pacciani (1993, p. 161), Todini (1998, p. 341), and Sitterson et al. (2017). Many rainfall-runoff modeling software such as MIKESHE, HSPF

(Hydrological Simulation Program Fortran), SWAT (Soil Water Assessment Tool), HEC-HMS, and so on (Sitterson et al., 2017) are available. Each of them is very unique and focused on specific applications. For example, HSPF is used in water quality modeling studies (Chandramouli et al., 2010, p. 520). MIKESHE integrates groundwater, evapotranspiration and soil moisture, surface runoff, and channel hydraulics and is widely used in water resources and environmental studies (McMichael and Hope, 2007, p. 245). Among them, HEC-HMS is very popularly used for flood modeling. HEC-HMS (Hydrologic Engineering Center – Hydrologic Modeling System) model is used very widely in the rainfall-runoff modeling process (Knebl et al., 2005, p. 325) because of its simplicity and modularized approach. It was developed by the U.S. Army Corps of Civil Engineers (HEC, 2016) and is available in the public domain. The HEC-HMS model is freely available for download from the U.S. Army Corps website (www.hec.usace.army.mil/software/hec-hms/downloads.aspx). In stage one, hydrologic modeling using the HEC-HMS model was done to simulate the watershed rainfall-runoff process (Chandramouli et al., 2019, p. 1). For developing the hydrologic model, geospatial data and field data are essential (Knebl et al., 2005, p. 325). Details of data needed and the processing steps are provided in Figure 6.3.

Datasets from the USGS (United States Geological Survey), USDA (United States Department of Agriculture), and USEPA (the United States

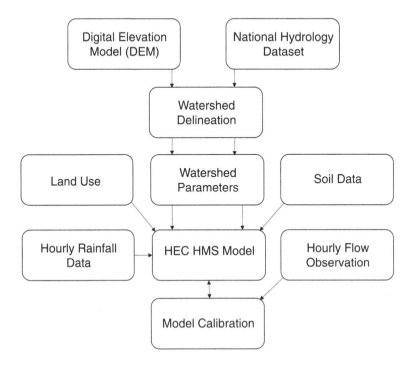

Figure 6.3 Initial data preparation.

Table 6.1 Data Sources Used for the Model Development

Data Type	Agency	Link
Digital elevation model (DEM)	USGS	www.usgs.gov/core-science-systems/ngp/tnm -delivery (through national map data delivery)
Watershed data	USGS	www.usgs.gov/core-science-systems/ngp/tnm -delivery (through national map data delivery)
National hydrography dataset	US EPA BASINS	Through BASINS software data download
Land use data	USGS	www.usgs.gov/core-science-systems/science -analytics-and-synthesis/gap/science/land -cover-data-download?qt-science_center _objects=0#qt-science_center_objects
Soil data	USDA	

Environmental Protection Agency) BASINS (Better Assessment Science Integrating Point and Non-point Sources) were downloaded and documented systematically for developing the HEC-HMS and HEC-RAS (Hydrologic Engineering Center – River Analysis System) models (Table 6.1). Necessary geoprocessing steps were completed using GIS (geographic information system) software initially to obtain watershed system data to set up the basic model. Hart Ditch Little Calumet River watershed was subdivided into nine sub-watersheds during the delineation process (Figure 6.4). Basic hydrologic modeling details such as routing, initial loss, Φ-index, base flow, and unit hydrograph are presented in Table 6.2. While delineating the watershed, a flow observation site was also used as a point of interest to set up a sub-watershed outlet. After the initial model development, using observed flow data, calibration was taken up to fine-tune the model to mimic a real-world system (Halwathura and Najim, 2013, p. 155). By properly setting up sub-watersheds during HEC-HMS model development, flow hydrographs were generated at different nodes of the watershed during the simulation. A well-calibrated model should be able to predict the time to peak and flow peak (Figure 6.5) close to that of field observations (Madsen et al., 2002, p. 48; McMillan et al., 2010, p. 1270; Yapo et al., 1996, p. 23). For calibrating the hydrology model, multiple storms were identified from the recent past and flow observations were downloaded from appropriate websites. An hourly time scale was used in the calibration process. Different storms were used for validation. Hydrographs generated from the calibrated HEC-HMS model became the input to the second stage hydraulic model.

6.2.3 Hydraulic Model Development

After satisfactory hydrology calibration of the HEC-HMS model, in the next stage, a hydraulic model was developed with HEC-RAS software for

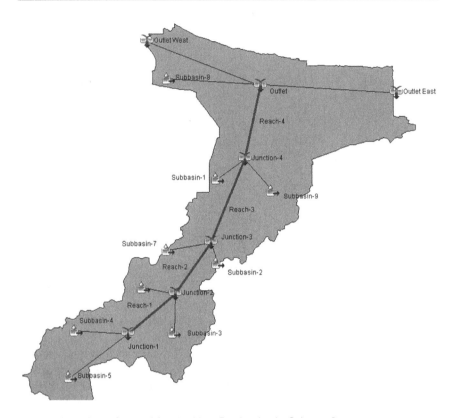

Figure 6.4 HEC-HMS model for the Hart Ditch – Little Calumet River system.

achieving flood modeling and plain mapping (Tate and Maidment, 1999). This model uses the finite difference method to solve Saint Venant Equations (HEC-RAS, 2016, Hicks and Peacock, 2005, p. 159). In addition to HEC-RAS, other software such as Mike 11 are also popularly used in the field for flood modeling a one-dimensional (1D) platform (Pappenberger et al., 2005, p. 46). The HEC-RAS 1D hydraulic model simulates flow through the channel and helps the user to identify the flood stages (HEC-RAS, 2016, Pappenberger et al., 2005, p. 46). The most recent version of the conventional HEC-RAS at the time of writing is version 5.06, available for free download from the U.S. Army Corps site. It can be used for both steady and unsteady flow simulations. The HEC-RAS modeling scheme is data intensive. For the model development, one needs cross-sections at different nodes of the channel, longitudinal section, Manning's "n" at different cross-sections, as well as flow hydrograph generated by hydrologic modeling (Lee et al., 2006, p. 319; Chandramouli et al., 2019, p. 1). Initially, reaches were constructed for the considered watershed using a background map (Figure 6.6) in the geometric editor of HEC-RAS. Cross-sectional surveys were done to

Table 6.2 Hart Ditch – Little Calumet River HEC-HMS Model Basic Details

Hydrologic System Component	Details
Watershed area	94.01 sq. miles. At USGS calibration gage site (USGS 05536190 Hart Ditch, Munster, Indiana), watershed area is 70.7 sq. miles drainage.
Number of sub-watersheds	9
Routing used	Muskingum method. X = 0.30 to 0.38 for different reaches after calibration. K = 0.3 to 1.1 hour for different reaches after calibration.
Transform – unit hydrograph	Snyder's UH Standard. Parameters calibrated. Peaking coefficient range – 0.1 to 0.2 for different sub-watersheds. Lag time range – 0.1 to 2.6 hour for different sub-watersheds.
Loss – initial loss and Φ-Index	Initial loss calibrated – 0.1 to 0.2 in. for different sub-watersheds. Φ-Index Calibrated – 0.2 to 0.31 in. per hour for different sub-watersheds.
Baseflow – recession	Calibrated values: Initial discharge (cfs/sq. mile) = 0.5 to 0.54 for different sub-watersheds. Recession constant = 0.79 Ratio to peak = 0.10 to 0.15 for different sub-watersheds.

set up cross-section profiles at different locations in each reach to develop an HEC-RAS model (Figures 6.6, 6.7, and 6.8). Steps involved in HEC-RAS model development are presented in Figure 6.9. Boundary conditions such as normal depth (friction slope) and unsteady flow data (hydrograph obtained from the HEC-HMS model) were included in the model in the next step. Using the observed flow and stage values, the HEC-RAS model was calibrated. After calibration, Manning's n at different locations varied

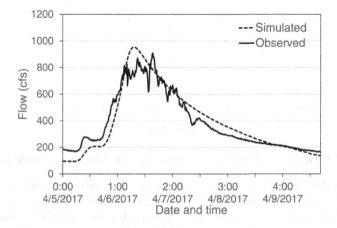

Figure 6.5 Flow hydrograph generated by HEC-HMS model for Hart Ditch, Munster, Indiana.

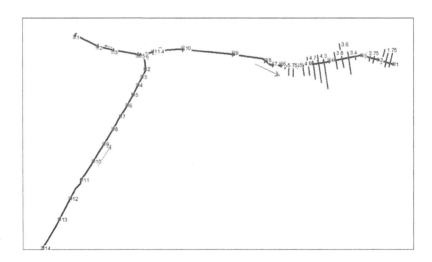

Figure 6.6 HEC-RAS model geometric editor for the Hart Ditch – Little Calumet River system.

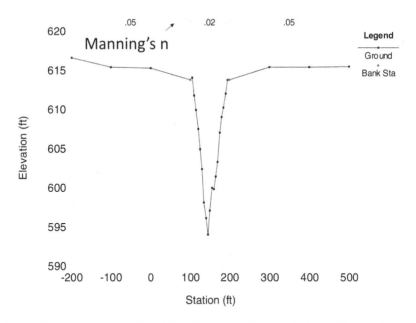

Figure 6.7 Cross section profile of Hart Ditch near Fran Lin Parkway, Munster, Indiana.

between 0.02 and 0.05. Uncertainties in handling different parameters during HEC-RAS calibration were very well presented in Pappenberger et al. (2005, p. 46). Multiple storms were used to calibrate and validate the model.

After successful calibration, results from the HEC-RAS model were captured by conducting several simulations using different site conditions to

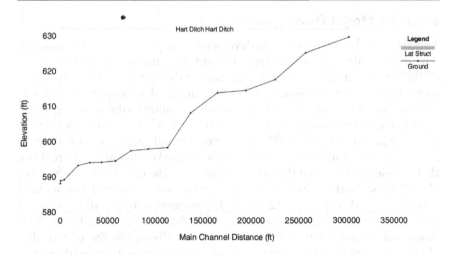

Figure 6.8 Longitudinal section profile for Hart Ditch, Indiana.

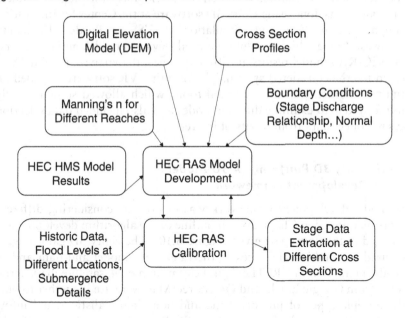

Figure 6.9 HEC RAS modeling steps.

develop the user-friendly flood module for student education. Simulation trials included the river without the levees, the river with different levee heights, with sluice gate three-fourths closed, half-closed, and fully open. Stages with respect to time at different cross-sections were used as the input to the virtual 3D model. After completing the hydraulic model calibration, stages with respect to time at different cross-sections were captured from the HEC-RAS model.

6.2.4 3D Model Development   •

The 3D virtual environment was developed to help students understand the boundary conditions, slow changes of land use pattern and their influence on flood peaks, modeling strategies, planning the remedial alternatives, and understanding their influences. Virtual reality models are powerful visualization tools, and they are used successfully in multiple educational applications in civil engineering (Dinis et al., 2017, p. 1683; Bell and Fogler, 1998, p. 1; Teegavarapu et al., 2007). Immersive virtual reality in these cases is defined as a computer interface that strives to make simulations so realistic that the users believe that they are experiencing "the real thing." Sampaio et al. (2011) used virtual reality techniques in the construction of a cavity wall and a bridge. Messner et al. (2003, p. 1) presented virtual reality tools that can be used in engineering education. Chandramouli et al. (2011, p. 934) developed virtual reality lab modules for visualizing the use of remedial alternatives developed for preventing solute transport with groundwater.

The interactive virtual reality software development for the flood education module was done using Unity 3D software at the Center for Innovation through Visualization and Simulation (CIVS) at Purdue University Northwest. Using a high-resolution digital elevation model, aerial imagery, and HEC-RAS model results at different nodes, the virtual 3D model was created for the considered system. The resulting VR software included an intuitive interface and informational tools, which allowed students to fly into the system to examine different nodes and also included data selection tools to collect data from points of interest.

6.2.4.1 Unity 3D Platform – A 3D Model Development Framework

The model development framework was evolved by considering different factors mentioned in Chapter 5. An outline of the algorithm development is presented in the flow chart given (Figure 6.10). The Unity 3D game engine was used to combine all pieces of the interactive model and data and to provide interaction in VR. The digital elevation model, satellite map, aerial photos from Google Earth, and OpenStreetMap were used together to provide multiple types of photorealistic and non-photorealistic visual information in the virtual environment. HEC-RAS stage results in different cross-sections were used for showing the flooding in the system.

6.2.4.2 Implementation

The Unity 3D platform provides many tools and accessories to create an interactive model. For developing the 3D model, the digital elevation model was used to capture the terrain for the considered watershed. Aerial photos were overlaid on the terrain as textures with proper georeferencing

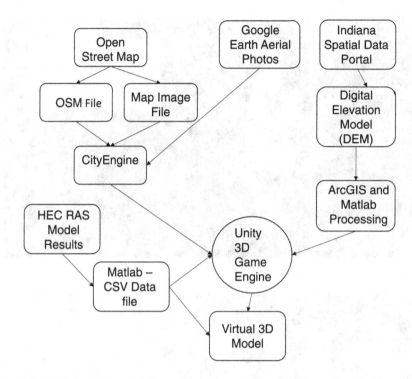

Figure 6.10 3D model development scheme.

steps to create the photorealistic 3D map view within a virtual environment. OpenStreetMap data downloaded from the public domain was overlaid above these layers. Every cross-section location used in the HEC-RAS model was located in the virtual model in the next step. Over the aerial photographs, homes in the neighborhood were constructed using Unity 3D tools. Using the water stage information at each cross-section, a water layer was generated. This generation gets terminated when the ground level becomes more than the water stage level. In between two cross-sections, the water stages were interpolated in the Y and Z direction. Using keys one can navigate through the virtual system. Using keys Q, E, D, A, S, W, and X one can move up, down, right, left, backward, and forward, respectively.

To facilitate the user exploration of data changing over time, a scroll bar was introduced. Using the time scroll bar, the user can visualize the changes in the system at different time steps. In the virtual environment, users can fly to different locations and change altitude to see details of each location or an overview of the flooding and surrounding areas. Apart from that, each cross-section across the creek was also listed for the users as interactable buttons. When a user clicks a cross-section, the user automatically flies to that location in the virtual model. When a user is at a cross-section,

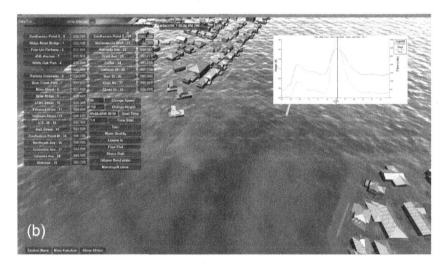

Figure 6.11 Virtual 3D model for Little Calumet River – Hart Ditch system.

they are provided with additional information and features, such as the ability to view a hydrograph of the storm at the current location (Figures 6.11a and b).

When the user moves the time steps using the time scroll bar, at that cross-section as well as in between sections, they can fly and examine the flooding scenarios in the neighborhood. Users can activate the 100-year flood plain (Figure 6.11c) and see how much it varies from the existing flooding at different locations. When the user drags the scroll bar, the user

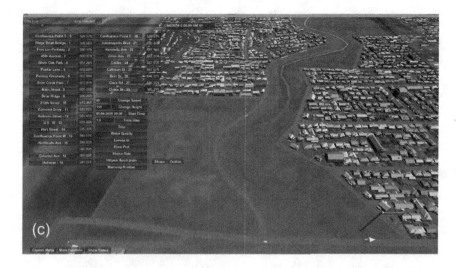

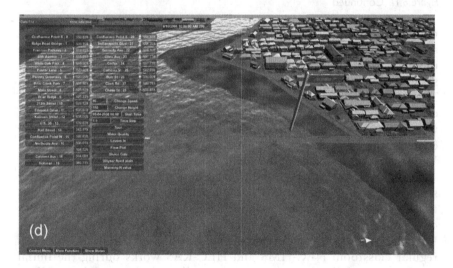

Figure 6.11 Continued

can see the status of flooding together with the hydrograph. Another tool allows the user to measure the flooding width (extent). When the user flies through the system, the location coordinates and altitude at which they are flying are also displayed. An on-screen compass indicates the north to help users keep track of their orientation as they fly through the virtual environment.

The Little Calumet River system also has a newly constructed levee system. In the virtual environment, the levee system is also included in the

Figure 6.11 Continued

model. Using different simulations made with the HEC-RAS model, with and without the levees (Figures 6.11d and 6.11e), options were given to the users to check the benefits of the levee system. Users can toggle the levee system to see the flooding extent with and without the levee system in place. They can also change the levee height and see the benefits derived from each case. Similarly, the software also has options to operate the sluice gate, which controls the flow in the Little Calumet River. Using the sluice gate controls, the influence of gate operations can also be studied by the users. Various options available in the application are shown in Figure 6.12.

The application can be used as a lab module in relevant courses for students. The usefulness of 3D-based learning was also evaluated by this tool in multiple programs at Purdue University Northwest, University of Kentucky, University of District of Columbia, and Florida Atlantic University. A baseline data was collected without a 3D lab module with regular classroom HEC-HMS and HEC-RAS works during the initial year. From the following year, data were collected after completing HEC-HMS and HEC-RAS labs as a pre-test (without 3D experience). With the same cohort of students, after a virtual 3D lab in CIVS, post-tests were done (with 3D experience). With 1D-modeling effort, students need to build a mental model to interpret results. By providing 3D experience in a user-friendly environment, students can comprehend things more easily. Based on this hypothesis, it was expected to have improvements in post-test performance. By collecting data from different universities, results were compared and analyzed (NSF DUE Grant #1245883). Pre- and post-tests were used to evaluate student learning within the VR application (Chandramouli et al., 2016a, p. 1).

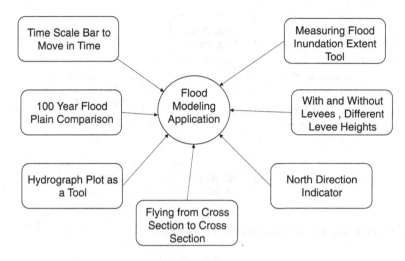

Figure 6.12 Flood model application – options.

6.3 VIRTUAL REALITY MODEL FOR GROUNDWATER CONTAMINATION MODELING

6.3.1 Introduction

Subsurface water constitutes a major part of the freshwater resource in the world (Masters and Ela, 2008). Groundwater represents by far the largest source of unfrozen freshwater on earth, which can be extracted easily and cheaply. In many regions around the world, it serves as the major source of drinking water as well as helping to meet industrial and agricultural requirements in addition to domestic uses (Todd and Mays, 2004). With an ever-growing need for freshwater, the risks concerned with groundwater are also increasing. As groundwater is utilized in the day to day life, the contamination of groundwater is mainly induced by mankind. Prominent contaminants to groundwater systems are chemical spills, gasoline spills, oil spills, leakage from the underground gasoline storage tank, and failing septic systems (Wagner and Shamir, 1992, p. 1233; Piver et al., 1998, p. 475; Li et al., 2007, p. 173). The students need to understand and visualize the problems related to groundwater contamination movement so that they will be able to handle the issue and develop suitable remedial alternatives.

6.3.2 Groundwater Modeling

As the groundwater contamination transport occurs underground, it is very difficult to develop a laboratory experiment with which students can visualize the problem. Even though there are few demonstrative models of groundwater available on the market, they are basically for understanding

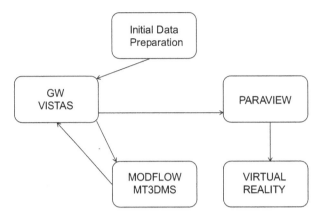

Figure 6.13 Groundwater model development scheme.

groundwater flow patterns only and not for chemical contamination movement. Due to the availability of powerful computing, several groundwater movement-based software such as MODFLOW (McDonald and Harbaugh, 1988) and solute transport models such as MODFLOWT (Duffield, 1996) and MT3DMS (Zheng and Wang, 1999) are available for users. MODFLOW (McDonald and Harbaugh, 1988) is a public domain software developed by the USGS. MODFLOW simulates three-dimensional (3D) groundwater flow using a finite difference solution.

Interfacing software called GWVistas (Rumbaugh and Rumbaugh, 2017) was used in this study for developing groundwater models (Figure 6.13). This software has many advantages such as facilitating finite difference mesh generation, editing data input using matrix worksheet, exporting data from different working platforms, and importing the results of groundwater software to different user-friendly formats. The software also provides flexibility to switch options between popular groundwater modeling software such as MODFLOW, MODFLOWT, MT3D, and MODPATH (Chandramouli et al., 2007, p. 1). Further, it also facilitates capturing results with respect to space and time in the model domain.

6.3.3 Case Study

For this research, a well-established benchmark problem presented in the MT3DMS manual was considered as a case study with slight changes. The demonstrative regional groundwater model developed was a quasi 3D model, 2000 ft long, 3000 ft wide, and 100 ft deep. Using GWVistas, finite difference grids were constructed for this region by using 168 rows and 108 columns. In the vertical direction, eight layers were considered. Other important aquifer and solute properties used to define the system are shown in Table 6.3.

Table 6.3 Basic Properties Used for the Case Study

Property	Value
Hydraulic conductivity (silt and clay)	50 ft/day
Hydraulic conductivity (sand and gravel)	70 ft/day
Hydraulic conductivity (fine sand)	250 ft/day
Porosity	0.3
Recharge rate	5 in/year
Distribution coefficient (K_d)	0.176 cm^3/g
Aquifer bulk density	1.7 g/cm^3
Longitudinal dispersivity	10 ft
Transverse dispersivity	2 ft

The finite grid mesh was developed in the area of interest with 50×50 ft size. The grid spacing was gradually changed toward the boundary. The contamination leak was at the center of the model domain (area of interest), where the maximum concentration is around 200 ppb. At the spill site immediately after the spill, 80% of the contaminant concentration was located in vertical layer 3 and the remaining 20% was in layer 2. All the other layers were clean at time $t = 0$. From east to west, the initial head of groundwater reduces with a gradient of 2.7×10^{-2} and it reduces with a gradient of 2.3×10^{-2} from north to south.

After developing the demonstrative model with chemical contamination plume, the base model run was performed without any remedial alternatives in place using MODFLOW, MT3DMS software, and GWVistas platform (Figure 6.13). Multiple simulations were conducted by changing boundary conditions with different remedial measures such as chemical barriers (Blowes and Ptacek, 1992, p. 214) and pump and treat (with different combinations of locations), and the results were documented.

6.3.4 Three-Dimensional Model Development in the Virtual Reality Environment

A virtual reality environment was used to create a realistic view of the groundwater medium and responds to the user's gesture and provides friendly interactivity to the user. Based on the interaction, it modifies the virtual world and permits the user to walk through the system and recognize the changing scenarios.

After completing the model runs for each case, the results were exported from the GWVistas software using the INP file format. Before this, several other options with groundwater modeling software were explored. The

INP option facilitates the data extraction for each finite difference grid cell of the quasi three-dimensional groundwater model for every temporal and spatial domain. This file was opened in a three-dimensional viewing software called Paraview (Ahrens et al., 2005). Paraview also facilitates the initial preview of the results in a three-dimensional setting. Several tasks such as color schemes, legend options, revising aspect ratio, surface treatment of the model, and soil grain introduction were performed with this software. Feedback from student users was documented and incorporated to refine the modules.

In this project, the VR software handled large amounts of data, around 3 GB for each case. Interacting with large datasets in real-time poses several challenges, and the task was handled by exploring different options. The large quantity caused low efficiency in processing and low frames per second (FPS) in displaying on the 3D VR system. The team tried its best to create a well-explained, highly visible, and friendly virtual underground environment. Each of the layers in the ground was colored based on the property of the soil considered. Then the team parsed the data to remove the invisible and irrelevant inner surfaces and simplified unimportant objects in the model (Figure 6.14). Further, to provide flexibility, care had been taken to separate and re-use parts of the model that remained unchanged at each time step from the data that was changing to obtain high efficiency in FPS (Chandramouli et al., 2011, p. 934).

This virtual 3D application (Burdea and Coiffet, 1993) helps to visualize the contamination movement through the groundwater system with respect to space and time. This application is a very powerful and versatile tool to

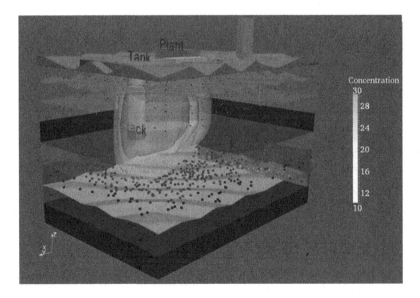

Figure 6.14 3D visualization of groundwater contamination scenario.

help the students to understand the complicated process which takes place underground. Students can view and fly through the underground scenario in the virtual environment and move around to see and explore the data in a more intuitive way than what is possible in physical labs. They can select different simulation options and examine each remedial measure performance such as deploying a chemical barrier or using wells. They can visualize the contaminant spread with respect to time. Using data picking tools at a given location, students can capture the contaminant concentrations with respect to time. Users can study the influence of different boundary conditions and remedial alternatives and collect data at select points with and without remedial measures and examine the benefits derived. They can also analyze the influence of boundary conditions. Both the lab modules are used in civil engineering labs regularly at Purdue University Northwest at CIVS center. In addition to 3D virtual reality, these applications can also be used in web-based environments and 2D format.

6.4 CONCLUSIONS

The future of VR in classroom teaching is inevitable. With technology growth, many such study tools can be implemented and used in both classroom and virtual learning settings. Models can be improved easily with feedback after use. Virtual reality platform can provide a hands-on experience to engage students and help them better understand complex systems and be better prepared for situations that they may encounter in the real world.

REFERENCES

Ahrens, J., B. Geveci, C. Law (2005). *Paraview: An End-User Tool for Large Data Visualization*. The Visualization Handbook, 717.

Bell, J. T., H. S. Fogler (1995). The investigation and application of virtual reality as an educational tool. *ASEE, Annual Conference*, Anaheim, CA.

Bell, J. T., H. S. Fogler (1998). Virtual reality in chemical engineering education. *ASEE, North Central Section Meeting*, University of Detroit Mercy, Detroit, MI.

Blowes, D. W. and Ptacek, C. J., (1992). Geochemical remediation of groundwater by permeable reactive walls: Removal of chromate by reaction with iron-bearing solids. *Proceedings of the Subsurface Restoration Conference*, 214–216, Dallas, Texas.

Burdea, G. C., P. Coiffet (1993). *Virtual Reality* Technology, 2nd Edition. John Wiley & Sons, IEEE Press, New York, USA.

Chandramouli, C. V., A. Galarza, E. Laviolatte, S. Pollard, A. Tracy (2019). Evaluating the new high flow channel performance in Little Calumet River System, Northwest Indiana. ASCE EWRI Congress 2019, Pittsburgh, OH. doi:10.1061/9780784482353.003

Chandramouli, C. V., E. Hixon, C. Q. Zhou, J. Moreland, J. Wang, Z. Xiong, R. Teegavarapu, P. K. Behera, J. F. Fox (2016b, June). Evaluating the usefulness of virtual 3-D lab modules developed for a flooding system in student learning. Presented at 2016 ASEE Annual Conference & Exposition, New Orleans, Louisiana. doi:10.18260/p.26770

Chandramouli, C. V., Z. Ziong, M. J. Wang, J. Moreland, C. Q. Zhou, R. S. V. Teegavarappu, J. Fox, P. Behera, E. Hixon (2016a). Virtual 3D modeling for little calumet river flood inundation studies. ASCE EWRI Congress, May 2016, West Palm Beach, FL. doi:10.1061/9780784479841.011

Chandramouli, V., M. Narayana, V. Duruvai, S. Guo, J. Moreland, V. Merwade, C. Q. Zhou, A. Song, Y. Hu, F. Zhang (2011). Generalized visualization modules for solute transport in groundwater. EWRI ASCE Conference, May, 934–942, Palm Springs, CA.

Chandramouli, V., M. Narayana, R. S. V. Teegavarappu (2010). Developing fecal TMDLs for watershed management using fuzzy based approach. Watershed Management Conference, *EWRI, ASCE*, August 23, 520–528, Madison, Wisconsin.

Chandramouli, V., L. Ormsbee, J. Kopp (2007). Land acquisition study at paducah gaseous diffusion plant site using MODFLOWT modeling. ASCE EWRI Congress, May 15–19, Tamba, Florida, USA. doi:10.1061/40927(243)187

Dinis, F. M., A. S. Guimarães, B. R. Carvalho, J. P. P. Martins (2017). Virtual and augmented reality game-based applications to civil engineering education. *2017 IEEE Global Engineering Education Conference (EDUCON)*, Athens, 1683–1688.

Duffield, G. M. (1996). *MODFLOWT: A Modular Three-Dimensional Groundwater Flow and Transport Model*. HydroSOLVE, Inc. and GeoTrans, Inc., Sterling, Virginia.

Franchini, M., M. Pacciani (1993). Comparative analysis of several conceptual rainfall-runoff models. *Journal of Hydrology*, 122 (1–4), 161–219. doi:10.1016/0022-1694(91)90178-K

Fu, D., B. Wu, J. Moreland, G. Chen, S. Ren, C. Q. Zhou (2009). CFD simulations and VR visualization for process design and optimization. *Proceedings of the Inaugural US-EU-China Thermophysics Conference*, UECTC-RE T5-S6-0298, May.

Halwatura, D., M. M. M. Najim (2013). Application of the HEC-HMS model for runoff simulation in a tropical catchment. *Environmental Modelling & Software*, 46, 155–162 doi:10.1016/j.envsoft.2013.03.006

HEC-RAS in the Keelung River during the 2001 Nari typhoon. *Journal of Hydraulic Engineering*. ASCE, 2006; 132(3), 319–323.

Hicks, F E., T. Peacock (2005). Suitability of HEC-RAS for flood forecasting. *Canadian Water Resources Journal / Revue Canadienne des Ressources Hydriques*, 30(2), 159–174, doi:10.4296/cwrj3002159

https://www.lrc.usace.army.mil/Missions/Civil-Works-Projects/Little-Calumet-River/

Hydrologic Engineering Center (2016). *HEC-HMS: Hydrologic Modeling System*. U.S. Army Corps of Engineers, Davis, CA.

Kneb, M. R.., Z. L. Yang, K. Hutchison, D. R. Maidment (2005). Regional scale flood modeling using NEXRAD rainfall, GIS, and HEC-HMS/RAS:

A case study for the San Antonio River Basin Summer 2002 storm event. *Journal of Environmental Management*, 75(4), 325–336. doi:10.1016/j.jenvman.2004.11.024

Krishen, P., C. Watson, E. Douglas, A. Gontz, J. Lee, Y. Tain (2008). Coastal flooding in the northeastern US due to climate change. *Mitigation and Adaptation Strategy Planning of Global Change*, 13, Springer, 437–451.

Lee, K. T., Y. H. Ho, Y. J. Chyan (2006). Bridge blockage and overbank flow simulations using HEC-RAS in the Keelung River during the 2001 Nari typhoon. *ASCE Journal of Hydraulic Engineering*, 132(3), 319–323.

Li, J., G. H. Huang, G. Zeng, I. Maqsood, Y. Huang (2007). An integrated fuzzy-stochastic modeling approach for risk assessment of groundwater contamination. *Journal of Environmental Management*, 82, 173–188.

Madsen, H., G. Wilson, H. C. Ammentorp (2002). Comparison of different automated strategies for calibration of rainfall-runoff models. *Journal of Hydrology*, 261/1-4, 48–59.

Martino, G. D., N. Fontana, G. Marini, V. P. Singh (2013). Variability and trend in seasonal precipitation in the continental United States. *Journal of Hydrologic Engineering, ASCE*, 18. doi:10.1061/(ASCE)HE.1943-5584.0000677.

Masters, G. M., W. P. Ela (2008). *Introduction to Environmental Engineering and Science*, 3rd Edition. Pearson, New Jersey, USA.

McDonald, M. G., A. W. Harbaugh, (1988). A modular three dimensional finite-difference groundwater flow model. *Techniques of Water Resources Investigations*, Book 6, Chapter A1, United States Geological Survey, Reston, VA.

McMichael, C. E., A. S. Hope (2007), Predicting stream flow response to fire-induced landcover change: Implications of parameter uncertainty in the MIKE SHE model. *Journal of Environmental Management*, 84(3), 245–256.

McMillan, H., J. Freer, F. Pappenberger, T. Krueger, M. Clark (2010). Impacts of uncertain river flow data on rainfall-runoff model calibration and discharge predictions. *Hydrological Process.*, 24, 1270–1284. doi:10.1002/hyp.7587

Messner, J. I., S. C. M. Yerrapathruni, A. J. Baratta, V. E. Whisker (2003). Using virtual reality to improve construction engineering education. *Proceedings of the 2003 American Society for Engineering Education Annual Conference & Exposition*. Washington, DC.

Min, S. K., X. Zhang, F. W. Zweirs, G. C. Hegerl (2011). Human contribution to more-intense precipitation extremes. *Nature*, 378–381. doi:10.1038/nature 09763.

Pappenberger, F., K. Beven, M. Horritt, S. Blazkovac (2005). Uncertainty in the calibration of effective roughness parameters in HEC-RAS using inundation and downstream level observations. *Journal of Hydrology*, 302(1–4), 46–69. doi:10.1016/j.jhydrol.2004.06.036

Piver, W. T., L. A. Duval, J. A. Schreifer (1998). Evaluating health risks from ground-water contaminants. *Journal of Environmental Engineering*, 124, 475–478.

Poff, N. L. (2002). Ecological response to and management of increased flooding caused by climate change. The Royal Society, 360, 1488–1510.

Rumbaugh, J. O., D. B. Rumbaugh (2017). *Groundwater Vistas, User Manual.* C.S.I Environmental Inc, PA.

Sampio, A. Z., C. O. Cruz, O. P. Martins (2011). Didactic models in civil engineering education: Virtual simulation of construction works. *Virtual Simulation of Construction Works*, Prof. Jae-Jin Kim, InTech. ISBN: 978-953-307-518-1.

Sitterson, J., C. Kinghtes, R. Parmar, K. Wolfe, M. Muche, B. Avant (2017), *An Overview of Rainfall-Runoff Model Types, USEPA Report EPA/600/R-14/152*, September. Georgia, USA.

Tate, E. C., D. R. Maidment. (1999). *Floodplain Mapping Using HEC-RAS and ArcView GIS*. University of Texas, Austin.

Teegavarapu, R. S. V., P. Scarlatos, Y. Kaner (2007). A pilot study on catastrophic flood scenario animation for a region in South Florida. SFWMD Technical Report, *Center for Inter-Modal Transportation Safety and Security*, 38.

Thakur, B., R. Parajuli, A. Kalra, S. Ahmad, R. Gupta (2017). Coupling HEC-RAS and HEC-HMS in precipitation runoff modelling and evaluating flood plain inundation map. ASCE EWRI Congress, 240–251, Sacramento, California.

Todd, D. K., L. W. Mays (2004). *Groundwater Hydrology*, 3rd Edition. John Wiley and Sons, New Jersey, USA.

Todini, E. (1998). Rainfall-runoff modeling — Past, present and future. *Journal of Hydrology*, 100(1–3), 341–352.

Vergara, D., M. P. Rubio, M. Lorenzo (2017). On the design of virtual reality learning environments in engineering. *Multimodal Technologies Interact*, 1(2), 11. doi:10.3390/mti1020011.

Wagner, J. M., U. H. R. Shamir (1992). Nemati groundwater quality management under uncertainty: Stochastic programming approaches and the value of information. *Water Resources Research*, 28, 1233–1246.

Yapo, P. O., H. V. Gupta, S. Sorooshian (1996). Automatic calibration of conceptual rainfall-runoff models: Sensitivity to calibration data. *Journal of Hydrology*, 181/1-4, 23–48.

Zheng, C., C. P. Wang (1999). *MTEDMS, Documentation and User Guide*. Headquarters US Army Corps of Engineers.

Index

Printed in the United States
by Baker & Taylor Publisher Services